Treatment of Cancers by MEDICINAL PLANTS

The Author

Prof. Dr. Govind Pandey is probably "only person in Madhya Pradesh and alone veterinarian in India with maximum academic qualifications" (20 degrees/diplomas/certificates). His 'Biography' is included in the famous directory, book, "Who's Who in the World 2011" (28th edition, America). Having more than 34 years of experience in 'Research/Teaching/ Administration/Extension', Prof. Pandey graduated (BVSc & AH) from College of Veterinary Science and AH (JNKVV), Jabalpur in 1978, and obtained MVSc and AH and PhD (Hons.) in Pharmacology and Toxicology from the same college in 1980 and 1990, respectively. He has also secured MBA, LLB, LLM, MA (Soc.), MA (Hin.), MA (Eng.), MA (Pol.), Acharya (Astr.), PGDPA and LSG (Pub. Ad.), PGDCA, Sahitya Ratna, Ayurved Ratna, MDEH (Electro-Hom.), DNHE (Nut. & Hlt.), AIT (Art. Insem.), and PGPHT (Poul. Sci.). Presently, he is doing DSc. He is working on different aspects of Pharmacology and Toxicology (including Drug Nanotechnology, Fishery Science and other areas of Life Sciences), and Hindi literature. He has also keen interest in HRM, Political Science, Sociology, Public Administration, Law and Astrology.

Prof. Pandey has the unique distinction of working with 'excellence' in various government institutes/offices, non-government organizations and autonomous bodies. During his career started from 7th August 1980, Dr. Pandey worked in several capacities like "Veterinary Assistant Surgeon/ Lecturer/Veterinary Surgeon/Senior Veterinary Surgeon/Drawing Disbursing Officer (DDO)" under AH Department, Govt. of MP; "Chief Executive Officer/Block Development Officer/DDO" of some Janapad Panchayats under PRD Department, Govt. of MP; and "Assistant Professor/ Professor/Principal Scientist and Head" of Pharmacology in Pharmacy colleges. Besides, he served as "Deputy Director/Associate Professor/Senior Scientist" of Research Services, Nanaji Deshmukh Veterinary Science University (NDVSU), Jabalpur from 20th April, 2012 till he resumed the post of "Professor/Principal Scientist and Head", Department of Pharmacology and Toxicology, College of Veterinary Science and AH, Rewa (NDVSU, Jabalpur) on 26th November, 2012.

For great contribution in science/research/Hindi literature, Dr. Pandey has received '30 Awards/Honours/Fellowships/Recognitions'. His PhD work on 'indigenous hepatogenic drugs' was adjudged as the outstanding research work, for which he received "Sri Ram Lal Agrawal National Award" (1991; Indian Herbs, Saharanpur, UP). His other prestigious national awards in research include "Award for Excellence in Research" (Academic Brilliance Award-2014; Education Expo TV, Noida) and "Senior Research Fellowship" (1986; ICAR, New Delhi), besides many "Best Paper Awards". He has also been honoured with '3 Fellowship Titles': "Fellow- Academy of Sciences for Animal Welfare (FASAW)", "Fellow- Society of Life Sciences (FSLSc)" and "Fellow-International Science Congress Association (FISCA)". Prof. Pandey has supervised/guided/co-guided several PhD/PG/UG research scholars and carried out many projects. He has investigated "some antihepatotoxic and anticancer herbal drugs, oestrogen induced cancer model, and paracetamol induced hepatotoxic model". The research on 'oestrogen' induced experimental cancer and its treatment was successfully carried out for the first time in Asia, which was widely acclaimed by the eminent personalities, scientific societies and media.

Dr. Pandey has authored '11 Books/Manuals' and '3 Book Chapters' and edited '2 Books/ Manuals in science/research, besides many project reports/dissertations. He has published more than '235 Papers' and presented over '40 Papers' in conferences. In Hindi literature, Dr. Govind Pandey 'Prem' has published '5 Books', edited '1 Book', edited many magazines and published/ broadcasted several poems/lyrics/dramas/stories through different media. Dr. Pandey is the "Life Member/Office-Bearer" of about 25 scientific, professional, literary and cultural organizations/ associations/societies/journals. He has also been used as "Chairperson/Judge/Consultant/Resource Person" in many conferences/committees/programmes of different government and non-government organizations/associations/societies. He has been the "Editor/Mentor/Editorial Board Member/ Reviewer" of many journals/papers/books/magazines. He has delivered several speeches in a number of platforms. Besides, he was "Captain of Badminton", "Sergeant of NCC", "Literary Secretary" and "Hostel Prefect" in College of Veterinary Science and AH, Jabalpur; and has qualified "NCC B and C Certificates", and "2 years' Course of National Service Scheme".

Treatment of Cancers by
MEDICINAL PLANTS

Dr. Govind Pandey

BVSc and AH, MVSc & AH, PhD Hon. (Pharmacol.),
DSc (std.), LLB, LLM, MBA, MA (Soc.), MA (Hin.), MA (Eng.), MA (Pol.),
Acharya (Jyotish), PGDPA & LSG, PGDCA, AvR,
MDEH, SR, DNHE, AIT, PGPHT, FSLSc, FASAW, FISCA

Professor/Principal Scientist of Pharmacology and Toxicology,
The Nanaji Deshmukh Veterinary Science University,
Jabalpur, M.P., India

2015
Daya Publishing House®
A Division of
Astral International Pvt. Ltd.
New Delhi – 110 002

Cataloging in Publication Data--DK
Courtesy: D.K. Agencies (P) Ltd. <docinfo@dkagencies.com>

Pandey, Govind, author.
Treatment of cancers by medicinal plants / Dr. Govind Pandey.
pages ; cm
Includes bibliographical references (pages   ).

ISBN 9789351306313 (International Edition)

1. Cancer--Treatment. 2. Herbs--Therapeutic use. 3. Materia medica, Vegetable. 4. Medicinal plants.   I. Title.

DDC 616.99406   23

Published by   : **Daya Publishing House®**
A Division of
Astral International Pvt. Ltd.
– ISO 9001:2008 Certified Company –
4760-61/23, Ansari Road, Darya Ganj
New Delhi-110 002
Ph. 011-43549197, 23278134
E-mail: info@astralint.com
Website: www.astralint.com

Preface

"Cancer" is scourge afflicting mankind from the time immemorial, and is a group of more than one hundred diseases. It occurs when cells become abnormal and keep dividing and forming cells without control or order. Normally, the body cells divide to produce more cells only when the body needs them. If cells divide when new ones are not needed, they form a mass of excess tissue called *'tumour'*, that can be benign (not cancer) or malignant (cancer). In other words, *'cancer'* refers to an unrestrained proliferation and migration of cells. The cells in malignant tumour (cancer) can invade and damage nearby tissues and organs. The cancer cells can also break away from a malignant tumour and travel through the blood stream, or lymphatic system to form new tumour in other parts of the body. The spread of cancer is called *'metastasis'*. In fact, the cancer is major killer in most of the countries.

Several environmental factors, including air, water and industrial pollutants, environmental chemicals and radiations, etc. may cause various types of cancer. Certain specific substances or conditions can convert a normal cell into mutant cancerous type. *'Carcinogens'* are the physical and chemical agents that include the incidence of cancer in animal model and in human population. Many agents cause genetic damage and neoplastic transformation of cells. Chemical carcinogens, radiant energy, and oncogenic viruses and some other microbes can cause various cancers. There are also substances of variable or unclear chemical nature that can be carcinogens.

In spite of the spectacular advances made by Medical Science in the present century, treatment of cancer remains an enigma. Plants have been in use for treating different diseases from the prehistoric times and continue to be the source of more than 25 per cent of the present range of prescribed drugs. Since time immemorial, the medicinal plants and their products (or herbal drugs) have been used for treatment and prevention of different cancers. From the dawn of civilization, men have been utilizing the important biological properties of various plants for treatment of different diseases. In the indigenous system of medicine, plants have been used against many kinds of cancer, for over 2000 years all over the world. More than 1,000 plants have been found to exhibit significant anticancer activity. The earliest record of herbal treatment can be traced to ancient Chinese and Greek texts. Unani and Ayurvedic systems also used a large number of plants for the treatment of cancer. Nowadays, the worldwide research is going on to find out the effective anticancer drugs from plant origin since the allopathic system of medicine causes several side effects. It is now doubtless that plants are the most vital source of many compounds that possess significant therapeutic values for combating cancer.

From the time of Galen (ca. A.D. 180), the juice of woody nightshade (*Solanum dulcamara*) has been used to treat various tumours (including cancers), and warts. The main active tumour inhibitor has been identified as the steroidal alkaloid, glycoside β-solamarine. In folk medicine, even various lichens, *e.g.*, *Cladonia*, *Cetraria* and *Usnea* have a history of their application cancers. These plants are rich sources of usnic acid, a compound recognized for many years as antibacterial and antifungal, but only recently, as an antitumour compound. Similarly, many centuries ago, the Druids claimed that mistletoe (*Viscum album*) could be used to cure cancer. From the extract of *V. album*, 11 protein fractions with significant antitumour activity have been isolated. Mezeron (*Daphne mezereum*), despite its toxic properties, has also been recommended in many countries. The active antitumour constituents of this plant have been identified as a diterpene derivative, mezerin. However, prior to 1940s, the chief non-surgical treatment of cancer was X-ray and radium therapy, besides arsenicals and urethane, applied locally to destroy the inflicted tissues. During the 1940s, the radioisotopes, nitrogen mustards, sex hormones and antifolic acidic agents were developed. In recent years, efforts have been made to synthesize potential anticancer drugs. Consequently, hundreds of chemical variants of known classes of cancer therapeutic agents have been synthesized. It is recognized that a successful anticancerous drug should be the one which kills or incapacitates cancer cells without causing excessive damage to the normal cells. This criteria is difficult, or perhaps, impossible to attain, and that is why cancer patients suffers unpleasant side effects while undergoing treatments.

In this context, therefore, this reference/research book has been put forth to give the recent knowledge regarding different carcinogens leading to development of cancers, and their treatment by various medicinal plants/herbal drugs as discussed in different chapters of this book. This publication will definitely serve as resourceful referral to the scientists, oncologists, academicians and students of Life Sciences, including Veterinary, Medical, Ayurvedic and Pharmacy. Last but not the least, our sincere thanks and deep regards are devoted to Dr. A.S. Nanda (Hon'able Vice Chancellor), Dr. D.N. Srivastava (Ex-Professor and Head of Pharmacology and Toxicology), Dr. A.B. Shrivastav (Director, Centre for Wildlife Forensic and Health), Dr. S.P. Shukla (Dean, College of Veterinary Science & AH, Rewa), Dr. Y.P. Sahni (Director, Research Services), and Dr. Madhuri Sharma (Associate Professor of Fishery Sciences), all at NDVSU, Jabalpur; and to Dr. S.P. Pandey, Professor and Head of Pharmacology, Govt. NSCB Medical College, Jabalpur; and to Dr. Asha Khanna,

Professor and Head of Zoology and Biotechnology, St. Aloysius' College, Jabalpur, MP. All the authors/writers/editors/journals/books of the matters and photographs cited in this book are gratefully acknowledged; my heartfelt gratitude and sincere thanks would only be a token appreciation of what they deserve for completion of this book.

Dr. Govind Pandey

Contents

Chapter 1

An Overview on Cancer, Carcinogenesis and Carcinogens

Cancer

'*Cancer*' (malignant tumour) is a disease of human and other multicellular animals. It is an abnormal growth and proliferation of cells. It is a frightful disease as the patient suffers pain, disfigurement and loss of many physiological processes. Cancer is may be uncontrolled and incurable, and may occur at any time at any age in any part of the body. It is caused by a complex, poorly understood interplay of genetic and environmental factors. Being chronic in nature, it is difficult to cure cancer as the cells proliferate to their maximum. Cancer represents the largest cause of mortality in the world and claims over 6 million people each year (Madhuri 2008; Pandey, 2007; Pandey and Madhuri, 2006a, 2006b; Pandey *et al.*, 2013a and 2013b; Somkumar, 2003). Thus, cancer is an abnormal growth of cells in the body that leads to death. Cancer cells usually invade and destroy the normal cells. These cells born of an imbalance in the body and by correcting imbalance, the cancer may go away. Billions of dollars have been spent on cancer research and yet we still don't understand exactly what cancer is (Madhuri and Pandey, 2009a; Pandey *et al.*, 2013b).

Cancer is scourge afflicting mankind from the time immemorial, and is a group of more than one hundred diseases. It occurs when cells become abnormal and keep dividing and forming cells without control or order. All organs of the body are made up of cells; normally the cells divide to produce more cells only when the body needs them. If cells divide when new ones are not needed, they form a mass of excess tissue called '*tumour*', that can be benign (not cancer) or malignant (cancer). The cells in malignant tumour (cancer) can invade and damage nearby tissues and organs. The cancer cells can also break away from a malignant tumour and travel through the blood stream, or lymphatic system to form new tumour in other parts of body. The spread of cancer is called *'metastasis'*. In fact, the cancer is major killer in most of the countries. This disease is a growing public health menace that gives rise to more than 6 million new cases every year (Pandey *et al.*, 2013a; Rajan and Chezhiyan, 2002).

In other words, *'cancer'* refers to an unrestrained proliferation and migration of cells. A mass of cancer cell descended from a single parental cell is called *'tumour'*. The tumour posing no danger to life or health is called *'benign'* and if it does cause danger, it is called *'malignant'*. Some tumours continue to grow at their original site, while others stop growing and shrink. Some tumour cells break free of the original tumour, migrate to new site in the body and establish new tumour, that is referred to as *'metastasis'*. The cancer originating from the cells on organ liners and forms part of the skin is known as *'carcinoma'* (more than 80 per cent of cancers belong to this category). The cancer originating from the cells of connective tissues, bones or muscle tissues is called as *'sarcoma'*. The cancer originating from the glandular tissue is called as *'adenocarcinoma'*. However, the cancer which starts from the white blood cells (WBCs) is referred as *'leukemia'* (blood cancer); while the cancers formed from the non-neuronal cells of brain are known as *'glioma'* and *'astrocytoma'*. Over 200 types of cancer syndromes have been identified in humans, and almost all types of cell, tissue, organ and system are vulnerable to cancer development (Lakshmi Prabha *et al.*, 2002; Pandey *et al.*, 2013a).

Carcinogenesis

'Carcinogenesis' is that process which results in the appearance of lesions which meet the requirement of a cancer (Ferguson and Denny, 1999; Pandey *et al.*, 2013b). Carcinogens reach the somatic cell and directly or indirectly react with DNA of respective cell. By direct interaction, carcinogens damage DNA, either inhibit the activity of topoisomerase II, or break the DNA by poisoning the enzyme through epigenetic mode. The indirect interaction has no action with genetic material but still it can induce cancer, *e.g.*, peroxisome proliferate binds at appropriate receptor site in plasma membrane and modifies fatty acid metabolism, through which it can cause cancer (Ferguson and Denny, 1999). Irrespective of cancer types, following stepwise sequential processes lead to develop the cancer up to metastasis conditions (Pandey *et al.*, 2013a):

i. *Stage I (Initiation)*- Firstly, mutation occurs due to exogenous and/or endogenous factor.

ii. *Stage II (Promotion)*- Secondly, cell turnover, growth factors, hormones and non-getoxic chemicals are increased.

iii. *Stage III (Progression)*- Thirdly, genetic instability, mutation of normal oncogenes, suppressor gene loss and chromosomal aberration occur.

iv. *Stage IV (Metastasis)*- Fourthly, tumour invasion, activation of genes for angiogenesis, cell adhesion and proteolysis, etc. occur.

Usually, all types of cancer are preventable; however, the habitual and environmental conditions are the main aetiological factors for the development of many cancers. Certain specific substances or conditions can convert a normal cell into mutant cancerous type. These substances or conditions are as under (Lakshmi Prabha *et al.*, 2002):

a. The functional and metabolic status of a presumable cell to become a cancerous one in future, especially in the critical contribution of *'oxygen species'*.

b. The carcinogenic substances, which include the physical, chemical, alcoholic and non-alcoholic beverages like coffee, tea, tobacco and betel leaf.

c. The viral or other pathogenic infections activate the pathogenesis into oncogenes and cause cancer; or in some cases, genetic predisposition inactivates the tumour suppressor gene and induces cancer through mutations of hereditary type.

The normal cells require oxygen to live and when they are deprived of oxygen, they become cancerous within two-cell generations. Further supply of oxygen to these cells continues to maintain them as cancerous cells. These transformed cancerous cells consume glucose and utilize oxygen to convert it into lactic acid, instead of carbon dioxide (as the normal cells do). Due to this abnormal conversion, liver gets activated and converts lactic acid back to glucose. The cancerous cells once again utilize the reverted glucose, thus making normal metabolism more complicated. The changed metabolism affects normal body (noncancerous) cells, and they are deprived of sufficient amount of glucose, thus making regular cellular metabolism. Also, the routine function of liver is increased and overburdened mainly because of accumulated lactic acid. The cancerous growth causes liver to deviate from normal functioning which affects the entire metabolic pathway in the cancer patient (Pandey *et al.*, 2013a). Hence, *'oxygen'* in the form of *'free radicals'* (the molecules, which are highly reactive, they bind and destroy body components) affects normal cells and can cause cancer. The *'oxygen free radicals'* can also cause many diseases, including heart disease (Lakshmi Prabha *et al.*, 2002).

Carcinogens

'Carcinogen' is a substance which may cause the development of cancers that otherwise would not have appeared. An *'anticarcinogen'* in one or another way prevents the appearance of tumours which might be expected when a carcinogen is administered (Pandey *et al.*, 2013a). Carcinogens are physical and chemical agents that include incidence of cancer in humans and animals (Ferguson and Denny, 1999).

A large number of agents can cause genetic damage and neoplastic transformation of cells. Chemical carcinogens, radiant energy, and oncogenic viruses and some other microbes can induce different types of cancer. The radiant energy and some chemical carcinogens are the documented causes of cancer in humans, and evidence linking certain viruses to human cancer grows ever stronger. The carcinogen classifications deal with the groups of substances like aniline and homologs, chromates, dinitrotoluenes, arsenic and inorganic arsenic compounds, cadmium compounds, nickel compounds, crystalline forms of silica, beryllium and its compounds, organophosphorus compounds and pesticides. There are also substances of variable or unclear chemical makeup that may be carcinogens, some of which include coal tar pitch volatiles, coke oven emissions, diesel exhaust and tobacco smoke (Kumar *et al.*, 2006b; Madhuri, 2008; Pandey *et al.*, 2013a).

Chemicals, *e.g.*, dichloro-diphenyl-trichloroethane (DDT), kepone, methoxychlor and toxaphene (all insecticides), atrazine (a herbicide), polychlorinated biphenyls (PCBs) and diethylstilbestrol (DES) that possess oestrogenic activity and can act as endocrine disruptors leading to development of many cancers (Mukherjee *et al.*, 2006). Irradiation is the main cause of physical form of cancer in human skin. The actual mechanism behind the X-rays related cancer is chromosomal damage. The ultraviolet (UV) rays can also cause skin cancer wherever the autonomic recessive disorders and xerodermic conditions are predominant. In some cases, the skin overexposed to direct sunlight may become cancerous by DNA repair mechanism. All the physical carcinogens probably first attack the dermal system where the melanin pigment is very low. The malignant melanoma and other skin cancers are common in black skinned people, but particularly in fair-skinned cells (Harris *et al.*, 2001).

Some important classes of chemical carcinogens are as under (Kumar *et al.*, 2006b; Lakshmi Prabha *et al.*, 2002; Madhuri, 2008; Pandey *et al.*, 2013a):

A. Direct Acting Carcinogens

    i. Alkylating Agents: *e.g.*, β-propiolactone, dimethyl sulphate, diepoxybutane and anticancer drugs (such as cyclophosphamide, chlorambucil, nitrosoureas, etc.). Most of the drugs of this group consumed for various ailments bind with DNA, leading to development of cancer.

    ii. Acylating agents- *e.g.*, 1-acetyl-imidazole and dimethyl carbamyl chloride.

B. Indirect Acting Carcinogens (Procarcinogens)

    i. Polycyclic and heterocyclic aromatic hydrocarbons (PAHs and HAHs)- *e.g.*, benz(a)anthracene, benzo(a)pyrene [B(a)P], dibenz(a,h)anthracene, 3-methyl-cholanthrene and 7, 12- dimethyl-benz(a)anthracene (DMBA). Many of these are derived from cigarette smoke, car exhaust and fumes. They react with base pairs of cellular DNA, inflict changes and result in cancer, as in the case of lung cancer.

    ii. Aromatic amines, amides and azo dyes- *e.g.*, 2-naphthylamine (β-naphthylamine), 2-acetylaminoflucrene, dimethyl-aminoazobenzene (butter yellow) and benzidene. β-napthylamine and benzidene are the main active forms that react with liver cells and once excreted from liver, they reach urinary bladder to cause bladder cancer.

    iii. Nitrosamine- They are stored or tinned foods containing nitrites, which are converted into nitrates when ingested. Nitrates react with amines, causing cancer.

    iv. Natural plants and microbial products- *e.g.*, aflatoxin B_1 (one of the most important organic toxins produced by the fungus, *Aspergillus flavus*), griseofulvin, cycasin, safrole and betel nuts. The aflatoxin combines with hepatitis B virus and causes liver cancer.

v. **Others-** *e.g.,* vinyl chloride, nickel, chromium, insecticides, fungicides, some amides and PCBs.

There are other types of chemical carcinogens that enter into our body in various forms. Of these forms, tobacco (chewing tobacco and snuff), alcohol, coffee, a high fat containing diet and areca nut with betel leaf are important that can cause cancer if their concentrations exceed in the body. The optimum level of caffeine in the plasma is 30 µg. If this level increases, that would cause chromosomal aberration (Pandey *et al.,* 2013a).

Certain Views on Cancer Development

The high frequency radiation, *e.g.,* ionizing radiation and UV radiation may cause genetic damage, leading to cancer. The ionizing radiations include X-rays, γ-rays, cosmic rays and particles given off by radioactive materials, such as α and β particles, protons and neutrons. Radiant energy (UV rays of sunlight, ionizing electromagnetic or particulate radiation) induced the neoplasm (cancer) in both human beings and experimental animals. The ionized molecules are unstable and quickly undergo chemical changes, thereby forming the *'free radicals'* that can damage the DNA molecule or other molecules around it. The ionizing radiations can cause mutation in a cell's DNA, resulting into cancer. Thyroid gland and bone marrow are most sensitive to radiations; while kidney, bladder and ovary are least affected. Some forms of leukemia (bone marrow cancer) appear to be the most common radiation-induced cancers. The atomic blasts in Japan showed that high-dose radiation increases the risk of developing several cancers, including leukemia. Breast cancer can be developed in elderly women by a much more extensive use of personal computers (more than 3 hours a day), mobile telephones, TV sets and other household electrical appliances. Children are twice as sensitive as adults to leukemia-causing effects of radiation, and unborn children exposed to radiation in the uterus are even more sensitive. The breast cancer risk is more than twice as high as normal; however, the risk of developing lung cancer is 50 per cent higher and the risk for multiple myeloma is more than twice as high as in the general population. The radiation therapy at high doses can cause DNA mutations in cells that survive radiation, which may lead to a second primary cancer. Of all the types of non-ionizing radiations, only UV rays are cancer-causing agents. The most skin cancers are a direct result of sunlight exposure as sun is the major source of UV radiation (Pandey and Madhuri, 2010a).

Environmental hazards or contaminants may cause malnutrition, which ultimately causes many diseases, including cancer both in humans and animals. Malnutrition is directly responsible for 300,000 deaths per year worldwide in children younger than 5 years in developing countries. However, cancer is the second leading cause of human deaths in the world and claims over 6 millions every year. It may be caused due to incorrect diet, genetic predisposition or environmental factors. At least 35 per cent of all cancers worldwide are caused by incorrect diet (Pandey, 2009 and 2010). In the case of colon cancer, malnutrition may account for 80 per cent of the cancer cases. Hence, the food eaten must not only be nutritious but also it must be complete and clean without any contamination, otherwise the person eating the food would get ill even the food is nutritious. If the right food is not consumed in right amounts by a person, it results in malnutrition. The most important environmental or xenobiotic (foreign) chemicals causing malnutrition and cancer are PCB biphenyl congeners, pesticides, food-related mycotoxin and its derivatives, ultraviolet screen, camphor, some metals, fungicides, algicides, oestrogens, retinoids, pyrethroid insecticides, pentachlorophenol, β-hexachlorocyclohexane, etc. (Pandey, 2010).

Environmental and food contaminants, including therapeutic agents (e.g, antibiotics, antihypertensive drugs, etc.) and oestrogenic endocrine disruptors (xenoestrogens or xenobiotic chemicals) mediate oestrogens, leading to development of cancer. Oestrogens and their metabolites are involved in the cancer pathogenesis of breast, uterus, ovary, pituitary gland, testicle, liver, kidney and bone marrow. Xenoestrogens that can cause cancer include PCB congeners, pesticides, food-related mycotoxin zearalenone and its derivatives, UV screen, some metals, fungicides, algicides, oestrogens, retinoids, pyrethroid insecticides, pentachlorophenol, β-hexachlorocyclohexane, etc. Diet has influence on cancer development, and several compounds, either present as dietary components or contaminants or formed during food processing, may play a role in cancer risk. All these can adversely interfere with the physiological functions of oestrogens; mediate the hormonal responses by binding to oestrogen receptors; interact with steroid hormone binding proteins; inhibit oestrogen synthesis; and overall can interact with enzyme systems to modulate the oestrogen metabolism (Pandey and Madhuri, 2010b).

Many DNA and RNA viruses have been proved to be oncogenic (or carcinogenic) in a variety of animals, ranging from amphibia to primates, and evidence grows stronger that certain forms of human cancer are of viral origin. Several DNA viruses have been associated with the causation of cancer in animals. Adenoviruses cause tumours in laboratory animals, while bovine papilloma-virsues cause both types of tumours (neoplasm), *i.e.*, benign neoplasm and malignant neoplasm (cancer) in their hosts. Of the various human DNA viruses, human papilloma viruses (HPV), Hepatitis B virus (HBV), Epstein-Barr virus (EBV) and Kaposi sarcoma herpes virus have been reported to cause cancer in humans. The human 'T' cell leukemia virus type 1 (HTLV1) and Hepatitis C virus (HCV) of RNA type are also reported to cause cancer. HPV have been implicated in the development of many cancers, such as squamous cell carcinoma of cervix, oral cancer and laryngeal cancer. Bovine papillomavirus (BPV) is known to cause epithelial and mucosal tumours in cattle. HBV and HCV may be the primary cause of hepatocellular carcinoma (HCC). EBV (a member of herpes family) has been reported to cause lymphomas and nasopharyngeal carcinomas (Pandey and Madhuri, 2010c).

Pollution of rivers and streams with chemical contaminants has become most critical environmental problems. Fish living in a polluted water reservoir (in the vicinity of an oil refinery) use contaminated water to rinse their gills, resulting into deposition of PAHs in the fish body. Contamination of foodstuffs by heavy metals, *e.g.*, arsenic (As), cadmium (Cd), chromium (Cr), nickel (Ni) and lead (Pb) has posed a potential carcinogenic threat to humans. The As and Cd appear to be the most harmful to fish. Many cancers in fish appear to be the result of exposure to different environmental pollutants/chemicals (Madhuri *et al.*, 2012a). Several chemical compounds, environmental pollutants, radiant energy, and oncogenic viruses and some other microbes can cause genetic damage and neoplastic transformation of cells, leading to development of various cancers. Many cancers in fish appear to be the result of exposure to most of such environmental pollutants/chemical agents. The carcinogens can be exposed through aquarium water and liver cancer induction, in particular, provides best results with most of the carcinogens so far studied. High frequencies of liver and skin cancers in brown bullheads are associated with high concentrations of PAHs and some metals in the sediment. The HCC in sauger and walleye are associated with heavy loadings of extremely fine particulates, which were produced when copper *'stamp sands'* were reprocessed. The cancers in zebrafish are more common. The beluga whales are getting cancer, while those in the less-polluted water do not suffer from cancer (Pandey *et al.*, 2013b).

Most of the times, the chemical contaminants or environmental pollutants cause severe toxicity, leading to various types of cancer in fishes. The presence of chemical carcinogens, combined with fish tumour induction data and reports of tumours in the wild fish population have provided strong circumstantial evidence that chemical carcinogenesis is occurring at some sites. Liver has been found to be the most targeted organ for fish cancer. The liver (hepatic) and skin cancers have been reported in brown bullheads in association with PAHs and other contaminants. The cancers which have been reported in zebrafish are seminoma, HCC, HCA, adenoma of exocrine pancreas, intestinal adenocarcinoma, ultimobranchial neoplasm, thyroid neoplasm, spindle cell sarcoma and hemangioma. Diethylnitrosamine (DEN) is one of the most potent carcinogens. The most common type of interstitial cell neoplasm reported to occur in several fish species is hemangiopericytoma. Fish neoplasms/cancers in intrahepatic and extrahepatic abdominal regions, have been induced by methylazoxymethanol acetate (MAMA) and N-methyl-N'-nitro-N-nitrosoguanidine (MNNG). Epizootics of skin cancer/neoplasia have also been associated with chemical contaminants. Although the environmental contaminants are important factors in the development of hepatic and skin cancers in fishes, the causes may be multifactorial, thus more laboratory research identifying specific mechanisms is needed (Pandey *et al.*, 2013c).

Oestrogens are most commonly used as oral contraceptives (OCs) and hormonal replacement therapy (HRT) in women. They are also indicated for the treatment of various reproductive disorders in females and prostate cancer in males. At the same time, oestrogens are used for the treatment of different reproductive problems in dogs. In spite of its great usefulness, oestrogen is responsible for several detrimental effects, including cancer/carcinogenesis in humans as well as animals. Excessive and prolonged use of oestrogen has been reported to cause cancers of endometrium, cervix, vagina, ovary, liver and mammary gland in women. Ethinyl oestradiol (EO, a highly potent semisynthetic oestrogen) caused the uterine and ovarian cancer in rats. The evidences broadly conclude that oestrogen even at therapeutic dose may produce several hazards, including cancer. The extent and severity of tissue damage have been found to be dose and time dependent, suggesting that the higher dose with prolonged duration of oestrogen administration may cause more cytotoxic changes (Madhuri, 2008; Madhuri *et al.*, 2012d).

The cancerous effect of oestrogen (EO) at a specific dose and duration was observed on some gynecologic organs (uterus and ovary) of albino rat. On the 21st week, the uteri of EO (250, 500 and 750 µg/kg, orally, weekly for 20 weeks) treated rats revealed varying degrees of congestion, degeneration and necrosis in the mucosa and endometrial gland with infiltration of inflammatory cells. Rats treated with EO at the dose of 750 µg per kg, showed severely necrosed endometrial glands; the eosinophil infiltration was quite conspicuous. Proliferation of endometrium in the form of papillary projections was seen, indicating the hyperplasia of endometrium as cancerous growth. At places, the hyperplastic epithelium appeared to invade the basement membrane also. The ovaries of EO (250, 500 and 750 µg/kg, orally, weekly for 20 weeks) treated rats showed marked congestion with vascular wall thickening, fibrosis, degeneration and necrosis in follicular tissues. However, the rats treated with EO at the dose of 750 µg per kg, revealed papillary proliferation in surface epithelium and hyperplasia of follicular cells, which indicated the development of cancer. The extent and severity of uterine and ovarian damage were dose dependent, suggesting that at higher dose for prolonged period, oestrogen can cause gynecologic (uterus and ovary) cancer in women (Madhuri *et al.*, 2007b).

Uterine cytotoxicity leading to cancer has been produced after administration of EO in rats. In a study, EO was dosed at the rate of 750 µg per kg, orally, weekly for 8, 12, 16, 20 and 24 weeks in different groups of albino rats. On the 9th week, degeneration and necrosis with infiltration of inflammatory cells in the uterine tissues were noticed. On the 13th week, marked vascular congestion, epithelial necrosis and extensive fibrosis, resulting into compression of endometrial glands were seen. On the 17th week, endometrial glands were severely necrosed and conspicuous eosinophil infiltration was seen. On the 21st week, endometrial tissues showed more severe changes, including sloughing of epithelium with infiltration of inflammatory cells. Focal hyperplasia of endometrial lining in the form of papillary projections (as endometrioid mass) and fibrovascular connective tissues were seen. The hyperplastic epithelium appeared to invade the basement membrane also, indicating the cancerous growth. On the 25th week, hyperplastic and cancerous changes in the endometrial tissues appeared much more severe and extensive. The extent and severity of uterine damage were time dependent, as EO after prolonged period (20-24 weeks) caused severe changes (Madhuri, 2008).

Similar to production of uterine cancer by EO (estrogen), another study was conducted to induce an experimental model of ovarian adenocarcinoma (cancer) by administration of EO (@ 750 µg/kg, orally, weekly for 16, 20 and 24 weeks) in rat. The EO on the 17th week caused degeneration and necrosis of ovarian follicular tissues, thick banding leading to fibrosis, thickening of vascular wall, and presence of homogeneous mass in the lumen. On the 21st week, fibroplasia of interfollicular connective tissues, hyperplasia of follicular cells and papillary proliferation in surface epithelium were observed. On the 25th week, other malignant lesions such as hyperchromatosis, enlargement of nuclei and anisokaryosis, anisocytosis were also seen. These changes indicated the development of adenocarcinoma in the rat ovary.

The extent and severity of ovarian damage caused by EO were time dependent (Madhuri and Pandey, 2010b).

The carcinogenic effect of different doses EO was also observed in the ovaries of albino rats. The EO was administered to different groups of rats in doses of 250, 500 and 750 µg per kg body weight, orally, weekly for 16 and 20 weeks. On the 17th week, the ovarian tissues revealed marked congestion, fibrosis, degeneration and necrosis of follicular tissues. On the 21st week, these changes were more marked. In general, extensive fibrosis, thickening of blood vessel walls, follicular tissue degeneration and necrosis were noticed. At this period, EO (750 µg/kg) caused papillary proliferation in the surface epithelium and hyperplasia of follicular cells, indicative of cancer in the rat ovary. The extent and severity of ovarian damage were dose and time dependent, suggesting that EO at higher dose for prolonged period (750 µg/kg, orally, weekly for 20 weeks) may produce hyperplasia leading to cancer in ovary (Madhuri and Pandey, 2011).

Chapter 2

Causation and Development of Cancer

Cancer by Radiations

Many studies showed that several environmental factors, including air, water and industrial pollutants, environmental chemicals and radiations, etc. may cause various types of cancer (Kumar *et al.*, 2006b; Mukherjee *et al.*, 2006; Pandey, 2007; Pandey and Madhuri, 2010a).

Emission of energy from any source is called *'radiation'*. There is hardly any aspect of human welfare in which nuclear radiation does not play an important role. The use of radiation has become an integral part of modern life. Radiation is used for scientific purposes, medical reasons and for power and energy generation. High frequency radiations, *e.g.*, ionizing and UV radiations may cause genetic damage, leading to cancer (Kumar *et al.*, 2006b). Radiant energy, whether in the form of the UV rays of sunlight or as ionizing electromagnetic and particulate radiations, induces neoplasm (cancer) in both human beings and experimental animals. The ionizing radiations include X-rays, γ-rays, cosmic rays and particles given off by radioactive materials, *e.g.*, α and β particles, protons, and neutrons. The ionizing radiations consist of high-energy waves that are able to penetrate cells and can cause ionization. The ionized molecules are unstable and quickly undergo chemical changes, thereby forming free radicals that can damage the molecule or other molecules around it. One type of molecule that is sensitive to ionizing radiation is DNA. Thus, the ionizing radiations can lead to a mutation (change) in a cell's DNA, which could contribute to cancer, or to damage/death of cell. The amount of damage is related to the dose of radiation received by the cell. The organs differ in their sensitivity to the effects of radiation. Thyroid gland and bone marrow are most sensitive to radiations; while the kidney, bladder and ovary seem to be least affected. Some forms of leukemia (cancer of bone marrow) appear to be the most common radiation-induced cancers.

Ionizing radiation is an effective way to treat certain kinds of cancer (Sen, 2008). Cancer of thyroid gland follows closely but only in young person; while in intermediate category are the cancers of breast, lung and salivary gland (Kumar *et al.*, 2006b).

Effects of Radiations

Generally, the health is affected on the basis of amount and duration of radiation exposure. The health can be reduced or death may occur due to radiation exposure as described below (Bhatia, 2008; Pandey and Madhuri, 2010a):

  i. *Stochastic health effects-* These are associated with long-term, low-level (chronic) exposure to radiation. 'Stochastic' refers to likelihood that something will happen. Increased levels of exposure make these effects more likely to occur, but do not influence the type or severity of effect. Cancer is the primary health effect from radiation exposure. The radiations can cause mutations in DNA.

  ii. *Non-stochastic health effects-* These effects appear due to high levels of radiation exposure and become more severe as the exposure increases. The short-term, high-level exposure is called *'acute exposure'*. Many non-cancerous health effects of radiations are non-stochastic.

  iii. *Other long-term health effects-* Other than cancer, the most prominent long-term health effects are mutations (teratogenic and genetic). The genetic mutations can be passed from parent to child (offspring). The teratogenic mutations are caused by exposure of the fetus in uterus, and affect only the individual who was exposed. The teratogenic mutations can be smaller head or brain size with poorly formed eyes, abnormal slow growth and mental retardation.

Radiations Causing Cancer

The childhood leukemia was early connected to power-frequent magnetic fields of radiation already in the pioneering work by Wertheimer and Leeper (1979), and more recently Scandinavian scientists have identified an increased risk of malignant brain tumours, *e.g.*, astrocytoma and meningioma (Hardell *et al.*, 2005). Further, a clear association between adult cancers and FM radio broadcasting radiation has been noticed, both in time and location (Hallberg and Johansson, 2005). The breast cancer has developed in elderly women by a much more extensive use of personal computers (more than 3 hours a day), mobile telephones, television sets and other household electrical appliances. Among the elderly women who developed breast cancer in the first time frame, 20 per cent were regularly exposed to power-frequent fields. But in the more modern period, 51 per cent were so exposed, mainly through the use of personal computers. A significant influence of electromagnetic fields on the formation of epithelial mammary tumours has been observed (Beniashvili *et al.*, 2005). There is a strong association between the body-resonant non-ionizing radiation (FM radio, 100 MHz) and the existence of malignant melanoma of skin. Since this frequency range has a penetration depth of about 10 cm into the human body, there is a suspicion that resonant currents may affect the immune defense system also when it comes to beating cancer cells in lungs. Due to that, it is well motivated to study in detail how the presence and rate of lung cancer have changed in Sweden, United Kingdom and other countries as this new environmental factor was added (Pandey and Madhuri, 2010a).

No safe level or threshold of ionizing radiation has been stated. Even exposure to background radiation causes some cancers. Additional exposures cause additional risks. Hazards of exposure to some kinds of radiation were noted shortly after the discovery of X-ray in 1895. The skin reactions were observed in many people working with early X-ray generators and by 1902, the first radiation-caused cancer was reported in a skin sore. Within few years, a large number of such skin cancers were observed. The first report of leukemia in radiation workers appeared in 1911. Madam Marie Curie (discoverer of radium) and her daughter are believed to have died of radiation-caused leukemia. Since then, many studies have confirmed the cancer-causing effects of some types of radiation (Sen, 2008). The sun is the major source of UV radiation. There is ample evidence from epidemiologic studies that UV rays derived from the sun induce an increased incidence of basal cell and squamous cell carcinomas (the most common types of skin cancer), and possibly malignant melanoma of skin (Cleaver and Crowley, 2002). The UV rays have many effects on cells, inhibition of cell division, inactivation of enzymes and induction of mutation, leading to death of cells in sufficient dosage (Kumar *et al.*, 2006b). Further, the relationship between UV radiation and cancer of face, lower lip and other exposed areas (especially in fair-skinned individuals) is considered to be causal for a long period of time, particularly in people with outdoor occupation or rural residents (Sen, 2008).

Both electromagnetic (*viz.*, X-rays and γ-rays) and particulate (*viz.*, α and β particles, protons, and neutrons) radiations are carcinogenic. The evidence is so voluminous that few examples may suffice. Many pioneers of x-rays developed skin cancers. Miners of radioactive elements in the central Europe and Rocky mountain region of the United States have a tenfold increased incidence of lung cancers; most telling is the follow up of survivors of atomic bombs in Hiroshima and Nagasaki. Initially, there was increased incidence of leukemias, mainly acute and chronic myelocytic leukemia after an average latent period of about 7 years. Subsequently, the incidence of many solid tumours of breast, colon, thyroid and lung increased with longer latent periods. In humans, there is a hierarchy of vulnerability of different tissues to radiation induced cancers; most frequent are the leukemias, except chronic lymphocytic leukemia. An increased risk of breast cancer was seen decades later among women exposed during childhood to atomic bomb. An increase in thyroid cancer incidence was also seen in areas exposed to fallout from nuclear power plant accident in Chernobyl in 1986. Ionizing radiation therapy is an effective way to treat certain types of cancer. During radiation therapy, high doses of ionizing radiation are directed at the cancer, resulting in death of cancer cells. However, this can lead to DNA mutations in cells that survive radiation, which can eventually lead to another cancer, called a *'second primary cancer'*. An increase in second primary cancer in the area being irradiated has been seen in patients with many types of cancer following radiation therapy and/or chemotherapy. Treatment for Hodgkin disease (a type of lymphoma) often delivers lower radiation doses to many areas of body. These treated areas include large amounts of normal tissue. Patients with Hodgkin disease who are treated with radiation therapy are at an increased risk for developing second primary tumour. When considering radiation exposure from radiation therapy treatment, the benefits generally outweigh the risks. However, some combinations of radiation therapy and chemotherapy are more risky than others. Children treated with radiation therapy for hereditary retinoblastoma (a malignant eye tumour) are at an increased risk for developing a type of bone cancer called 'osteosarcoma'. Similarly, people who have nevoid basal cell carcinoma syndrome (a skin cancer) are at high risk for development of basal cell cancers in irradiated areas (Pandey and Madhuri, 2010a; Sen, 2008).

Cancer by Malnutritions and Environmental Hazards

Incidence of cancer is increasing worldwide and hence, cancer is the second leading cause of death. Indeed, cancer can be caused in one of the three ways: incorrect diet, genetic predisposition and via environment. At least 35 per cent of all cancers worldwide are caused by incorrect diet and in the case of colon cancer, diet may account for 80 per cent of the cases. When one adds alcohol and cigarettes to diet, the percentage may increase to 60 per cent. Genetic predisposition gives rise to 20 per cent cancer cases. Thus, most cancers are associated with a host of environmental carcinogens (Kumar *et al.*, 2006b; Pandey, 2007 and 2010).

Nutrition is closely associated with health. If a person eats right kind of foods in right amounts, that person will keep good health provided no other factors intervene. If the right food is not consumed in right amounts by a person it results in 'malnutrition' (*mal-noo-trish-un*). The malnutrition can result in either inadequate or excessive intake of food, which can ultimately lead to many diseases. Malnutrition is directly responsible for 300,000 deaths per year in children younger than 5 years in developing countries and contributes indirectly to over half the deaths in childhood worldwide. Globally, with 113.4 million children younger than 5 years affected as measured by low weight for age. The overwhelming majority of these children, 112.8 million, will live in developing countries with 70 per cent of these children in Asia, particularly the south-central region, and 26 per cent in Africa. An additional 165 million (29 per cent) children will have stunted length/height secondary to poor nutrition (Pandey, 2007 and 2010).

Environmental and food hazards or contaminants, including xenobiotics (environmental or foreign chemicals) like oestrogenic endocrine disruptors (xenoestrogens) and certain therapeutic agents (*e.g.*, antibiotics, antihypertensive drugs, etc.) may cause various types of cancer (Mukherjee *et al.*, 2006). The 'endocrine disruptors' influence the normal functions of oestrogens, thereby cause the cancers of several organs. Chemicals, industrial wastes, pesticides, OCs (or oestrogens), detergents, food additives and plastics, all are sources of environmental toxins and endocrine disruptors. The potential endocrine disrupters, *e.g.*, antibiotics, hormones, plasticizers and nonionic surfactants are becoming the priority pollutants. Most of these act as carcinogens (Mukherjee *et al.*, 2006; Sandhu, 2006), or they may contaminate the diet or food, thereby causing malnutrition (Pandey, 2007 and 2010).

Cancer and Malnutrition

'*Malnutrition*' refers to both undernutrition and overnutrition. '*Undernutrition*' means the inadequate intake of right food, while '*overnutrition*' means the excessive intake of right food. Deficiency of energy and protein may result in '*protein energy malnutrition*' (PEM), which is 'a pathological conditions arising from deficiency of protein and energy, and is commonly associated with infections'. This disorder has very serious consequences, leading to cancer and/or death. PEM is a major nutritional problem in India, and is widely prevalent among children (0-6 years) but is also observed as starvation in adolescents and adults, mostly lactating women, especially during famine period or other emergencies. If a person eats right kind of foods in right amounts that will keep good health provided no other factors intervene. The food eaten must not only be nutritious but also it must be complete and clean without any contamination, otherwise person eating the food would get ill even the food is nutritious (Pandey, 2007 and 2010).

India is facing a great problem of malnutrition, which is the single largest cause to the high rate of infant and child mortality and morbidity in India. The diet surveys showed that about 30 per cent of the families surveyed consumed inadequate amounts of food to provide the necessary nutritional requirements. It has also been established that an inadequate diet has contributed to high mortality and morbidity in the general population particularly in infants and pregnant women. The single major factor responsible for a wide prevalence of malnutrition in India is the '*poverty*'. About 50 per cent of our people live below poverty line and even after spending 80 per cent of their income on food, they can not have a balanced diet. There is an increased risk of malnutrition associated with chronic diseases like cancer, and diseases of intestinal tract, kidney and liver. The patients with these chronic diseases, especially cancer may lose weight rapidly and become susceptible to undernourishment because they can not absorb valuable vitamins, calories and iron. Malnutrition predisposes a person to infection; on the other hand, infection leads to malnutrition. This interrelationship and synergistic effect of malnutrition and infection often leads to a high incidence of cancer and/or deaths. The people who take contaminated or malnutritious diet are at greater risk of cancer. In relation to cancer, the malnutritional factors act as primary effectors in four situations: carcinogens in food articles, affected bioavailability of nutrients, nonnutritive dietary items and harmful contaminants. The malnutrition acts as carcinogen in many ways to produce cancer in humans as well as animals. Nutritional carcinogenesis may occur due to ingestion of toxins, nutritional deficiency or malnutrition, non-bioavailability of the micronutrients and inactivation of the metabolic enzymes (*i.e.*, mixed-function oxidases) present in the liver. Thus, the aetiological role of malnutrition preceding clinical cancer has been firmly established. People who are malnourished may be skinny or bloated. Their skin is pale, thick, dry, and bruises easily. Rashes and changes in pigmentation are common. Hair is thin and pulls out easily. Other symptoms of malnutrition include: anaemia, diarrhoea, loss of reflexes and lack of coordination, etc. All these signs and symptoms are correlated with those of many cancers (Pandey, 2007 and 2010).

Environmental Hazards Causing Cancer

Several studies reveal that environmental hazards or contaminants may either cause malnutrition or cancer. Many factors have a major effect on increasing the rates of oral, colon, lung and breast cancers. Some factors include increased infections, more use of pesticides, low consumption of fruits and vegetables, increased consumption of alcohol and red meat, more smoking, high industrial pollution, more exposure to sun, decreased physical activity and high occupational exposures. The major portion of chemicals to which humans are exposed is naturally occurring, that are carcinogenic at large doses. Almost every fruit or vegetable contains natural carcinogenic pesticides. Many natural chemicals are ingested as carcinogens from cooking food like roasted coffee. They are caffeic acid, catechol, DDT, furfural and B(a)P, etc. Food additives (*e.g.*, allyl isothiocyanate, alcohol, butylated hydroxyanisole and saccharin), mycotoxins (*e.g.*, aflatoxin and hepatitis B virus) and synthetic contaminants (*e.g.*, PCBs and tetyrachlorodibenzo-p-dioxin) are also mutagenic and carcinogenic (Gold *et al.*, 1997). Aryl hydrocarbons, *e.g.*, dioxins, PCBs and PAHs bind at the cellular aryl hydrocarbon receptor (AhR), leading to toxicity and carcinogenicity (Kawanishi *et al.*, 2003).

Humans are exposed to dioxins (belong to a group of halogenated aromatic hydrocarbons) mainly through contaminated foods. A dioxin compound, 2,3,7,8-tetrachlorodibenzo-para-dioxin (TCDD) is the most toxic congener. The TCDD, B(a)P and PCBs can activate AhR, which subsequently induce expression of cytochrome P4501A1 (CYP1A1) and P4501B1 (CYP1B1) enzymes. The oxidative stress caused by induction of cytochrome P450s is one of the toxic effects of dioxin. Pesticides like insecticides (*viz.*, carbaryl, carbosulfan, DDT, methoxychlor, dieldrin, endosulphan, sumithrin, organophosphorus compounds like malathion and parathion, etc.) and herbicides (*viz.*, alachlor, metolachlor, quarternary ammonium, atrazine, pentachlorophenol, etc.) have become a part of environmental contaminants due to their widespread use in agriculture and disease control. Many pesticides are immunotoxic and suppress the cell-mediated immunity. The pesticides like DDT and alachlor can behave like endogenous oestrogen and function to suppress apoptosis in estrogen receptor (ER)-positive human breast cancer. Some pesticides and related chemicals act as carcinogens. These xenobiotics adversely affect the lymphocyte function, and increase the oxidative stress and lipid peroxidation in various tissues. Studies indicate a close association between process of westernization and an increase in breast cancer incidence. Western countries showed higher breast cancer risk due to environmental influence. The lifestyles, including oral or dietary habits are responsible for incidence of many cancers. Diet with many compounds, either as dietary components or contaminants, or formed during food processing, influences to develop cancer. Zearalenone (nonsteroidal compound produced by *Fusarium* fungus species) can contaminate dairy products and cereals (*e.g.*, barley, corn, maize, rice and wheat). Zearalenone is a potential promoter of breast tumurigenesis. After consumption, food mutagens undergo metabolic activation or detoxification by different endogenous enzymes. Most mutagens begin their adverse effects at the DNA level by forming DNA adducts with carcinogenic metabolites. N-nitro compounds, *viz.*, N-nitrosamines and nitrosamides (*e.g.*, N-nitrosureas) are a large group of chemicals that are linked with cancer pathogenesis. Humans are exposed to N-nitrosamines from a variety of foods and tobacco smoke. Some pesticides like atrazine can be converted into genotoxic N-nitrosamines (*e.g.*, N-nitrosoatrazine) in the environment or digestive system. PAHs are formed during incomplete combustion of coal, oil, gas, garbage and other organic substances such as tobacco and different food items. Leafy vegetables can be a significant source of PAHs in human diet. Usually, PAHs occur in lower amount in cigarette smoke; human exposure is predominantly from dietary sources. Metabolic activation of PAHs results in DNA binding products. Thus, westernized dietary habits, including smoking may increase the breast cancer risk. Many lipophilic carcinogens of tobacco smoke, *e.g.*, PAHs like B(a)P and N-nitrosamines can be stored in breast tissues. Some PAHs such as 7,12-dimethylbenz(a)athracene (DMBA) and dibenzo(a,l)pyrene are potent mammary carcinogens (Mukherjee *et al.*, 2006; Pandey, 2010).

Tobacco has unfortunately become a routine part of the personal environment in the world. In India, 40 per cent of total cancer cases are tobacco related. Almost one fourth of the India's population (more than 250 million people) consumes tobacco. More than 10 million children below 15 years are addicted to tobacco. As per the World Health Organization (WHO), cancers of head, neck, lung, throat, urinary bladder, kidney, uterus and colon are mostly caused by tobacco. *'Bidi'* is said to be more harmful than cigarette. The smoke tobacco contains more than 4,000 chemical compounds and at least 400 toxins, many of them are known carcinogens and irritants. Some airborne contaminants are PAHs, carbon monoxide and nitrogen dioxide, etc. Tobacco related chewing materials include pan (betal), pan masala, gutka, zarda, snuff, tobacco powder, areca nut and naswar, etc. These may increase the risk of oral cancer. *'Pan masala'* is unfortunately charged with PAHs, organochlorine pesticides (DDT and benzene hexachloride–BHC), narcotics, metals and minerals. About 15 per cent of oral and pharyngeal cancers may be attributed to dietary deficiencies or malnutrition. Similarly, fungi (*Candida albicans*), viruses (human papilloma virus, hepatitis B virus, human T cell lymphotrophic virus and human immunodeficiency virus), UV radiation, immunosuppression, certain occupations (farming and industries) and poor socio-economic status aid in the development of precancerous lesions or carcinoma (Sen, 2008; Pandey, 2010).

Oestrogenic endocrine disruptors or xenoestrogens are widely distributed in environment. The xenoestrogens interact with the binding pocket of the ER because they have chemical similarities to oestrogen (usually a phenolic A-ring). Insecticides (*e.g.*, sumithrin) and pentachlorophenol alter steroidsignaling pathways. Other man-made chemicals acting as endocrine disruptors are atrazine, DDT, oestrogens (*e.g.*, oestradiol and ethinyl oestradiol), hexachlorophene and toxaphene, etc. (Mukherjee *et al.*, 2006). All these can lead to the development of various cancers (Pandey, 2010).

Pathogenesis of Cancer by Xenoestrogens

The lack of cell-cell adhesion and increased migration are key characteristics of cancer cells. All cancers occur due to activation or mutation of oncogenes, or inactivation of suppressor genes. More than 20 tumour suppressor genes and 50 oncogenes have been identified and characterized. Most of these genes are involved in activation and detoxification of PAHs, suggesting a potential role of these compounds in carcinogenesis. According to the WHO, cancer is the second largest cause of mortality and morbidity in the world after heart disease. *'Cancer'* is a word that put an immediate emotional response, which has no relationship to rational thinking, depth of knowledge of the individual and role of individuals in the society. Such responses also occur in doctors, nurses, patients, families, friends of those who suffer from cancer, people who think they may have cancer, teachers, scientists, politicians, social workers and unskilled laboures, etc. Cancer rates continue to rise each year and it kills over 16 million people worldwide per year. However, in India the incidence of cancer may be as high as 3 million cases every year. In addition, more than 12,00,000 deaths occur every year from various types of cancer (Sen, 2008). As per the other reports (Madhuri and Pandey, 2009a), cancer is the second leading cause of death in America; colon cancer is the second most common cause of cancer deaths and prostate cancer, second to skin cancer is the most frequently diagnosed cancer among men in the USA; while breast cancer is the most common of cancer in women worldwide. With increase in longevity, it is going to be a problem even in India. Cancers affecting the digestive tract are among the most common of all the cancers associated with aging. The major causes of cancer are smoking, dietary imbalances, hormones and chronic infections leading to chronic inflammation. Furthermore, cancer is one of the most dreaded diseases, and is leading cause of death; and presently in the USA, cancer accounts for 25 per cent of all deaths in humans (Balachandran and Govindarajan, 2005). Other reports indicate that cancer kills annually about 3500 per million population around the world (Kathiresan *et al.*, 2006; Madhuri, 2008; Madhuri and Pandey, 2009a; Pandey and Madhuri, 2010b).

Oestrogens are most commonly prescribed drugs by far the two major uses are as a component of OCs and HRT in women. OCs used to control the birth, have influenced the lives of untold millions of women since they may cause cancer in humans as well as animals. Due to wide use of OCs, a remarkable increase in cases of liver cell adenoma (benign or noncancerous tumour of liver) and hepatocellular carcinoma (HCC) occur. There is evidence that long-term use of OCs may increase the risk of breast and cervical cancers. Many studies on animals demonstrate an increased incidence of various types of tumour with oestrogen, hence with the use of OCs. Vaginal, uterine and breast carcinomas have been caused by oestrogens (Loose and Stancel, 2006; Madhuri, 2008; Madhuri and Pandey, 2010b; Madhuri *et al.*, 2007a; Pandey and Madhuri, 2008a; Satoskar *et al.*, 2005). The association between oestrogen use and cancer has been known for several years. *'Oestrogen cancer hypothesis'* postulates that both exogenous and endogenous oestrogens and their metabolites play an aetiologic role in the pathogenesis of cancer; and oestrogen administration is followed by regularly reproducible tumours of breast, uterus, ovary, pituitary gland, testicle, kidney and bone marrow either in mouse, rat, rabbit, hamster or dog (Hertz, 1976). Oestrogen can play a tumour-promoting role in target organs. This hormone is responsible for many illnesses in women as well as men. Oestrogens are closely linked with the pathogenesis of breast cancer. This risk is increased with prolonged use of oestrogen; hence, early menarche and/or late menopause increase the risk. The age at first pregnancy, nulliparity, obesity, oestrogen replacement therapy or

OCs are also related with the increased risk of breast cancer. The British and American studies in women showed that oestrogen increased the risk of uterine (endometrial and cervical) cancer, and the women who begin menstruating early or who start menopause late, produce more oestrogen over their lifetimes and have a higher risk of breast cancer (Madhuri, 2008; Pandey and Madhuri, 2010b).

Natural oestrogens, *viz.*, oestrone and oestradiol are recognized carcinogens in rodents and humans. Oestradiol increases the incidence of tumours of pituitary gland, mammary gland, uterus, cervix, vagina, testes, lymphoid system or bone in various strains of rats and mice (Liehr, 2001). In December 2000, the USA Government's National Toxicology Programme added oestrogen to the list of known human carcinogens. It is disconcerting to think that a natural hormone (oestrogen) circulating in significant amounts through the bodies of about half of the world's population (women) is a carcinogen. The EO (a semisynthetic 17 *b*-oestradiol oestrogen) causes cytotoxicity, leading to cancer in uterus, ovary and liver after chronic dosing in female rats (Madhuri, 2008; Madhuri and Pandey, 2010b; Madhuri *et al.*, 2012d; Pandey and Madhuri, 2010b).

Carcinogenic Action of Oestrogens

Oestrogens bind to cytoplasmic ERs in oestrogen responsive tissues. The steroids-receptor complex then translocates to the nucleus, where it brings about changes in transcription. There are two types of ERs, *viz.*, ER α (present in female genital tract, breast, hypothalamus, endothelial cell and vascular smooth muscle) and ER β (present in prostate, ovary and brain). Oestrogens act by interacting with specific ERs in the cytoplasm of target cells; this ER complex then translocates to nucleus, where it attaches itself to the appropriate gene and mediates the transcription of relevant mRNA (Loose and Stancel, 2006; Madhuri, 2008; Satoskar *et al.*, 2005). Differential expression of ER- α or β during ovarian carcinogenesis, with overexpression of ER-α as compared to ER-β in cancer has been demonstrated (Cunat *et al.*, 2004). This differential expression in ER suggested that oestrogen-induced proteins may act as ovarian tumour-promoting agents. After oestrogen hormone binds to its receptors in a cell, it turns on hormone-responsive genes that promote DNA synthesis and cell proliferation. If a cell happens to have cancer-causing mutations, those cells will also proliferate and have a chance to grow into tumours. In carcinogenic oestrogen, the reactions produce intermediates capable of producing oxygen radicals that can damage the cell's fats, proteins and DNA. Unrepaired DNA damage can turn into a mutation, leading to cancer (Madhuri, 2008; Madhuri and Pandey, 2010a and 2010b; Madhuri *et al.*, 2009; Pandey and Madhuri, 2010b; Platt, 2005).

Oestrogen-induced carcinogenic mechanism has been fully explained (Liehr, 2001). Oestradiol may be converted to 4-hydroxyoestradiol or oestrone. If these catechol oestrogens are not detoxified by phase II enzyme activities, *e.g.*, catechol-O-methyltransferase (COMT), UDP glucuronosyl transferase or sulphotranferase, they may undergo metabolic redox cycling. Quinone or semiquinone intermediates in the redox cycle are free radicals, which generate more oxygen radicals and may induce various types of DNA damage. This genotoxicity may result in gene mutation and cell transformation. Thus, the tumours may develop from cells transformed by genotoxic action of oestrogen, which proliferate in response to an ER-mediated stimulus (Figure 1).

In the metabolism of xenobiotics, cytochrome P450 enzymes or monooxygenases catalyze hydroxylation reaction. This process occurs in two phases: cytochrome P450s are involved in phase I reaction, which sometimes convert biologically inactive compounds into active or toxic metabolites. Subsequently in phase II, products of phase I reaction are conjugated with many molecules, *e.g.*, glucuronic acid, sulphate, glutathione (GSH), acetyl or methyl groups, leading to excretion from the body. Several phase I and II enzymes are associated with oestrogen metabolism. Oestrogens are mainly hydroxylated by cytochrome P4501A1 (CYP1A1) and CYP1B1 into 2-hydroxyestrogens and 4-hydroxyestrogens, respectively. It is believed that 4-hydroxyestrogens may act as a carcinogen. Therefore, the carcinogenic mechanism of oestrogen has been explained. The metabolisms of 2- and 4-hydroxyestrogens (catechol oestrogens) are produced in a series of linked oxidation reactions that form the oxidative oestrogen metabolism pathway causing cancer. Such metabolism leads to formation of unstable semiquinone, which is an intermediate in both oxidation and reduction reactions, and can react with molecular oxygen to form superoxide radicals and quinone. The superoxide radicals may be reduced to hydrogen peroxide and then to hydroxyl radical in the presence of metal ions. In general, quinones can be conjugated with GSH by glutathione-S-transferase (GST), or it can form adducts with guanine and adenine base in DNA. The 2,3-quinone can bind stably to DNA; whereas, 3,4-quinone forms depurinating adducts with guanine and adenine which are lost from DNA by cleavage of the glucosidic bond leaving apurinic sites with

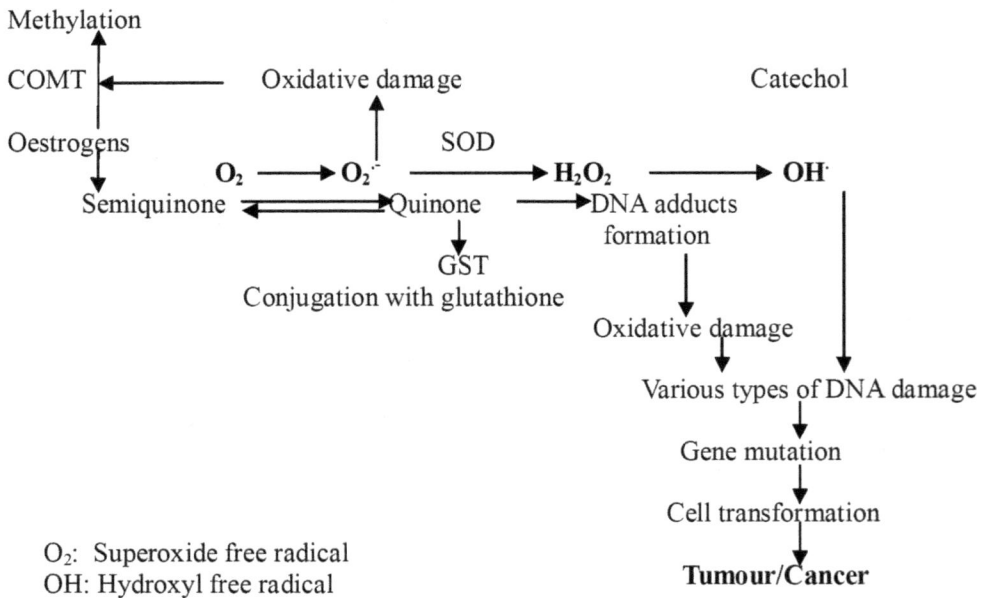

Methylation

COMT ← Oxidative damage   Catechol

Oestrogens   SOD

$O_2 \longrightarrow O_2^{\cdot-} \longrightarrow H_2O_2 \longrightarrow OH^{\cdot}$

Semiquinone ⇌ Quinone → DNA adducts formation

GST
Conjugation with glutathione

Oxidative damage

Various types of DNA damage

Gene mutation

Cell transformation

O_2: Superoxide free radical
OH: Hydroxyl free radical

Tumour/Cancer

Figure 1: Proposed Carcinogenic Action of Oestrogen [Liehr, 2001].

mutagenic potential. In addition, quinones and semiquinones undergo redox cycling which results in production of reactive oxygen species (ROS) that can cause oxidative damage to lipids, proteins and DNA. However, several phase II enzymes, *e.g.*, COMT, GST and superoxide dismutase (SOD) participate in catechol metabolism (Liehr, 2001; Madhuri, 2008; Mukherjee *et al.*, 2006; Pandey and Madhuri, 2010b; Figure 1).

Food and Environmental Contaminants as Carcinogens

Many factors have a major effect on increasing the rates of oral, colon, lung and breast cancers, etc. Some of these factors include increased infections, more use of pesticides, low consumption of fruits and vegetables, increased consumption of alcohol and red meat, more smoking, high industrial pollution, more exposure to sun, decreased physical activity and high occupational exposures. The major portion of chemicals to which humans are exposed is naturally occurring, that are carcinogenic at large doses. Almost every fruit and vegetable contains natural carcinogenic pesticides. Many natural chemicals are ingested as carcinogens from cooking food like roasted coffee. A diet free of naturally occurring carcinogenic chemicals is impossible. Food additives, mycotoxins and synthetic contaminants are also mutagenic and carcinogenic (Gold *et al.*, 1997). The aryl hydrocarbons, PCBs and PAHs bind to cellular AhR; and activation of intracellular signaling subsequent to AhR binding is highly correlated with the toxicity and carcinogenicity of these chemicals (Kawanishi *et al.*, 2003).

Humans are exposed to dioxins, mainly through contaminated foods. A dioxin compound, TCDD is the most toxic congener. TCDD, B(a)P and PCBs can activate AhR, which subsequently induce expression of CYP1A1 and CYP1B1 enzymes. Oestradiol is metabolized by these two enzymes, which activate B(a)P to reactive DNA-binding intermediates. There is evidence of cross-talk between ER α and AhR-mediated signaling in breast and endometrial cells, thus the oxidative stress due to induction of cytochrome P450s is one of the toxic effects of dioxin. Pesticides have become a part of environmental contaminants due to their widespread use in agriculture and disease control. Many pesticides are immunotoxic and found to suppress the cell- mediated immunity. Some of the selected pesticides are categorized as under: (a) insecticides such as carbamate (carbaryl, carbosulfan, primicarb, methicarb, etc.), organochlorine (DDT, methoxychlor, aldrin, dieldrin, endosulfan, heptachlor, etc.), organophosphorus compounds (dichlorovos, malathion, parathion, chlopyrifos, trichlorfon, etc.) and pyrethroid (bioallethrin, deltamethrin, sumithrin, etc.); (b) herbicides such as amide (isoxaben, pentachlor, alachlor, metolachlor, etc.), quarternary ammonium (paraquat, diquat, etc.), phenoxy (2,4-dichlorophenoxy acetic acid, 2,4-trichlorophenoxy acetic acid, etc.), trizine (atrazine, simazine, cyanazine, etc.) and unclassified group (oxadiazon, pentachlorophenol, etc.). The pesticides, such as DDT and alachlor can behave like endogenous oestrogen and function to suppress apoptosis in ER-positive human breast cancer. Some pesticides and related chemicals may act as carcinogens. These xenobiotics have adversely affected the lymphocyte function and increase oxidative stress and lipid peroxidation in various tissues metabolism. Besides reproduction-related factors, a westernized lifestyle is intimately associated with some particular food processing/cooking practices and dietary habits such as frequent use of fast foods. Some dairy products such as whole milk and different cheese containing high levels of saturated fat may increase cancer risk. Milk products also contain growth factor-I, which promotes breast cancer cell growth (Mukherjee *et al.*, 2006).

Zearalenone can contaminate dairy products and cereals. This compound may be a potential promoter of breast tumorigenesis. Nevertheless, there are some non-genotoxic carcinogens like chloropropanols (usually present in savoury foods). Furthermore, the N-nitro compounds are linked with the pathogenesis of cancer. Humans are exposed to N-nitrosamines from a variety of foods and tobacco smoke. Interestingly, some pesticides like atrazine can be converted into genotoxic N-nitrosamines environment or digestive system. B(a)P is the best PAH compound available from diet. In general, PAHs occur in lower amount in cigarette smoke; human exposure is predominantly from dietary sources. Metabolic activation of PAHs results in DNA binding products. Cooking meat and fish at high temperature produces heterocyclic amines. Generally, heterocyclic amines are formed from creatinine or creatine, amino acids and carbohydrates. Thus, the higher consumption of meat is probably associated with an increased risk of breast cancer.

The intake of alcohol can increase the cancer risk by enhancing the levels of circulating oestrogen (Mukherjee *et al.*, 2006). The food-related mycotoxin zearalenone and its derivatives such as α-zearalanol and β-zearalanol can bind to ER and exert oestrogenic action (Minervini *et al.*, 2005).

Tobacco is the only legally available consumer product that kills people when it is used as directed. WHO links tobacco to 25 cancers, *e.g.*, cancers of head, neck, lung, throat, urinary bladder, kidney, uterus and colon. In crowded environment, *e.g.*, bar and restaurant, second hand smoke (passive smoke) can produce six times the pollution of a busy highway. Smoking includes bidi, cigarette, chutta, cheroot, pipe, cigar, dhumti, hookas, kalke, chillum, etc. Tar (a mixture of many chemicals including formaldehyde, arsenic, N-nitrosamine, cyanide, benzopyrene, benzene, toluene and acrolein) and heavy metals (*e.g.*, As, Pb, Cd and Ni) may cause tobacco cancer. Many cancer promoters, initiators and accelerators as smoking agents are very potent carcinogens. These substances include nuclear aromatic hydrocarbons, chlorostilbenes, catechols, phenols and their substitutes, nitrosonornicotine, nicotine, cotinine nitrosamine and hydrazide. Chewing materials of human beings include pan (betal); *'Pan masala'* (a combination of areca nut, catechu, tobacco powder, aromatics, lime and brown or grey colouring agents); gutka; pan with tobacco (zarda, dokta or without tobacco); khaini; snuff; tobacco powder; areca nut; and naswar (a local preparation of Afghanistan commercially available as green tobacco powder mixed with calcium carbonate, calcium chloride, ammonium chloride, tobacco ignition residue, sand and other mineral substances). All these substances may increase the risk of oral cancer. Heavy and regular consumption of alcohol is a significant risk factor for oral and other cancers (Sen, 2008).

Endocrine Disruptors as Carcinogens

'Xenoestrogens or endocrine disruptors' are xenobiotic (environmental/foreign) chemicals that adversely interfere with the natural functions of hormones (Mukherjee

et al., 2006; Sandhu, 2006). The oestrogenic endocrine disruptors or xenoestrogen are widely distributed in the environment. Several chemicals like PCB congeners, pesticides (*e.g.*, dieldrin, endosulphan, pentachlorophenol), food-related mycotoxin zearalenone and its derivatives, ultraviolet screen, 3-(4-methylbenzylidene)-camphor and even some metals like cadmium can influence hormonal responses by binding to ER. The xenoestrogens interact with the binding pocket of the ER because they have chemical similarities to estrogen (usually a phenolic A-ring). Many xenobiotics can interact with the enzyme systems that metabolize oestrogens; and by this process they may modulate the endogenous metabolism. Numerous chemicals such as fungicides (*e.g.*, fenarimol, procymidone, vinclozolin) and algicides (*e.g.*, triphenyltin) inhibit steroid hormone synthesis. Synthetic compound like DES interfere the bio-availability and overall functions of oestrogen by interacting with steroid hormone binding proteins in the blood. The AhR or dioxin receptor has involvement with ER-mediated response pathways. Chemicals like retinoids, pyrethroid insecticides (*e.g.*, sumithrin), pentachlorophenol and β-hexachlorocyclohexane alter steroid signaling pathways. Other man-made chemicals which can act as endocrine disruptors (related with oestrogens) include atrazine, DDT, EO, hexachlorophene and toxaphene, etc. These can influence the oestrogen functions, such as adverse effects on release and excretion of hormones, disruption of regulatory feedback relationships between two endocrine organs, and modulation of non-genomic pathways (Mukherjee *et al.*, 2006; Pandey and Madhuri, 2010b).

Virus Induced Carcinogenesis

Many environmental factors, including air, water and industrial pollutants, environmental chemicals and radiations, etc. may cause various cancers (Madhuri, 2008; Pandey and Madhuri, 2010b and 2010c). The environmental factors may also be different microbes, including viruses. Many DNA and RNA viruses have been proved to be oncogenic in a variety of animals, ranging from amphibia to primates, and evidence shows that certain human cancers are of viral origin (Pandey and Madhuri, 2010c).

The EBV, a herpes virus, was discovered from a Burkitt lymphoma cell line in 1963 (Epstein, 2001). It was the first virus identified from a human neoplastic cell, followed by HPV, HBV, HCV, HTLV1 and human herpes virus type 8 (HHV8) (Uozaki and Fukayama, 2008). The RNA and DNA oncogenic viruses have made fundamental contributions to the areas of cancer research. Transforming retroviruses carry oncogenes derived from cellular genes that are involved in mitogenic signalling and growth control. Viruses are now accepted as bonafide aetiologic factors of human cancer; these include HBV, EBV, HPV, HTLV1 and HCV, and several candidate human cancer viruses. It has been estimated that 15 per cent of all human tumours (or neoplasms, cancers) worldwide are caused by viruses. The infectious nature of viruses distinguishes them from all other cancer-causing factors; cancer viruses establish long-term persistent infections in humans, with cancer an accidental side effect of viral replication strategies. Viruses are usually not complete carcinogens, and the known human cancer viruses display different roles in transformation. Most infected individuals do not develop cancer, although immunocompromised individuals are at elevated risk of viral-associated cancers (Butel, 2000; Pandey and Madhuri, 2010c).

The genomes of oncogenic DNA viruses integrate into and form stable associations with host cell genome. The virus is unstable to complete its replicative cycle because the viral genes essential for completion of replication are interrupted during integration of viral DNA. Thus, the virus can remain in a latent state for years. Those viral genes that are transcribed early in the viral life cycle (early genes) are important for transformation, and are expressed in transformed cells (Kumar *et al.*, 2006b). The 'T' antigen proteins encoded by DNA tumour virus early genes are involved in the transformation of normal cells to immortalized neoplastic cells that may or may not be tumorigenic in immunocompetent animals. Studies have been made of the tumorigenicity of DNA virus-transformed cells and the interactions of these cells *in vivo* and *in vitro* with immunologically nonspecific host effector cells, such as natural killer (NK) cells and macrophages. Results imply that 'T' proteins determine the capacity of transformed cells to induce tumours by governing the level of susceptibility that transformed cells express to destruction by such host cellular defenses. An *in vitro* system to study the carcinogen-induced amplification in simian virus 40 (SV40), transformed Chinese hamster (CO60) cells have been described. DNA amplification of helper-dependent parvovirus AAV (adeno-associated virus) can be induced by a variety of genotoxic agents in absence of co-infecting helper virus. The AAV origin/terminal repeat structure has been recognized by cellular DNA replicative machinery induced or modulated by carcinogen treatment in absence of parvoviral gene products (Pandey and Madhuri, 2010c).

Oncogenic DNA Viruses

Many DNA viruses have been associated with the causation of cancer in humans and animals. Adenoviruses can cause tumours in laboratory animals, while BPV cause benign as well as malignant neoplasms in their hosts (Kumar *et al.*, 2006b). BPV caused epithelial and mucosal tumours in cattle. Followings are the important oncogenic DNA viruses products (Pandey and Madhuri, 2010c):

1. Human Papilloma Viruses (HPV):

The HPV are small DNA tumour viruses, belonging to the family of Papovaviridae. About 70 genetically distinct types of HPV have been identified. HPV have been implicated in the development of many cancers, *e.g.*, squamous cell carcinoma of cervix, oral cancer and laryngeal cancer. The incidence of cervical carcinoma worldwide is estimated as high as 400,000 diagnosed per year. More than 90 per cent of high-grade cervical dysplasias and invasive cervical cancers have been associated with 10 to 15 high-risk HPV viral types. Epedemiologic studies suggest that carcinoma of cervix is caused by a sexually transmitted agent and HPV is the culprit. The DNA sequences of HPV 16 and 18 and less commonly, HPV 31, 33, 35 and 51 are found in about 85 per cent of invasive squamous cell cancers. In cancer, the viral DNA is usally intregrated into the host cell genome. The oncogenic potential of HPV 16 and 18 can be related to viral E6 and E7 early viral gene products, which act in conjunction to immortalise and transform cells (Kumar *et al.*, 2006b). The replication of DNA viruses is dependent on the replication machinery of the host cells, and E6 and E7 act to overcome the activity of cell cycle inhibitors. E6 and E7 enhance p53 degradation, causing a block in apoptosis and decresed activity of p21 cell cycle inhibitor. E7 associates with p21 and prevents its inhibition of the cyclin D/CDK4 complex; E7 can bind to RB, removing cell cycle restriction. The net effect of HPV E6 and E7 proteins is to block apoptosis and remove the restrains to cell proliferation (Helt and Galloway, 2003; Kumar *et al.*, 2006b; Figure 2).

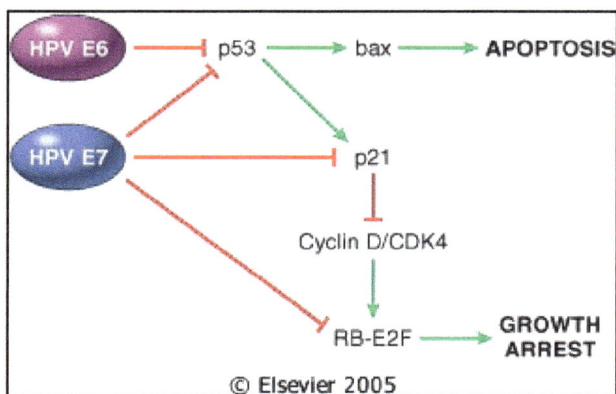

Figure 2: HPV Proteins E6 and E7 Acting on Cell Cycle [Helt and Galloway, 2003; Kumar *et al.*, 2006b].

2. Epstein-Barr Virus (EBV)

The EBV (a member of herpes family) has been found to cause the pathogenesis of four types of human tumours, *i.e.*, the African form of Burkitt lymphoma, B cell lymphoma in immunosupressed indivisuals, some cases of Hodgkins's lymphoma and nasopharyngeal carcinomas (Dolcetti and Masucci, 2003). EBV infects epithelial cells oropharynx and B lymphocytes. Within B lymphocytes, the linear genome of EBV circularises to form an episome in the cell nucleus. The infection of B cell latent, *i.e.*, there is no replicaion of virus and the cells are not killed, but the latently infected B cells are immortalized and acquire the ablity to propagate indefinitely *in vitro*. The molecular basis of B cell immortalization by EBV is complex. It appears that EBV serves as one factor in the multistep development of Burkitt lymphoma (Figure 3). More than 90 per cent of the world population is infected with EBV before adolescence, but EBV-associated malignant neoplasms develop in a limited number of patients in an endemic or non-endemic manner. Furthermore, EBV is associated with the transformation of various types of cells like lymphoid, dendritic, smooth muscle and epithelial cells. EBV-associated gastric carcinoma (GC) is the monoclonal growth of EBV-infected epithelial cells. EBV-associated GC is distributed worldwide and more than 90,000 patients are estimated to develop GC annually in association with EBV (10 per cent of total GC). It occurs in two forms in terms of the histological features, *i.e.*, lymphoepithelioma-like GC and ordinary type of GC. Both share characteristic clinicopathological features such as the preferential occurrence as multiple cancer

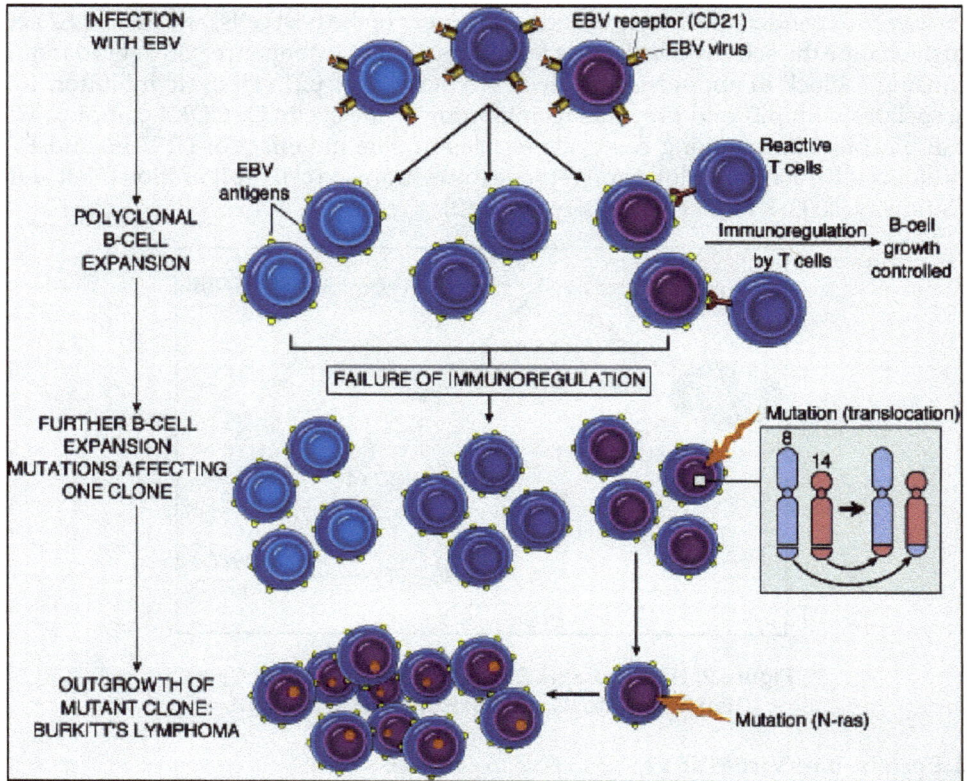

Figure 3: Development of Burkitt Lymphoma by EBV
[Dolcetti and Masucci, 2003].

and remnant stomach cancer. The EBV-associated GC shows gastric cell phenotype, resistance to apoptosis, and the production of immunomodulator molecules (Uozaki and Fukayama, 2008).

3. Hepatitis B Virus (HBV)

The HBV is a small enveloped DNA virus which primarily infects hepatocytes and causes acute and persistent liver disease. Chronic HBV infection is a major risk factor for the development of HCC. The HBV is endemic in countries of the Far East and Africa; these areas have the highest incidence of HCC. Studies in experimental animals also support a role for HBV in the development of liver cancer. In virtually all cases of HBV-related liver cancer, the viral DNA is integrated into the host cell genome, and as with HPV, the tumours are clonal with respect to these insertions (Kumar *et al.*, 2006b). The role of HBV in carcinogenesis appears to be complex, and may involve both direct and indirect mechanisms. Chronic liver inflammation and hepatic regeneration induced by cellular immune responses may favour the accumulation of genetic alterations. The HBV DNA may also disrupt or promote the expression of cellular genes, which are important in cell growth and differentiation. The rate of chromosomal alterations is significantly increased in HBV-related tumours compared with tumours associated with other risk factors. HBV might, therefore, play a role in enhancing genomic instability. The study also suggested that the chronic HBV infection triggers oncogenic pathways, thus playing a role beyond stimulation of host immune responses and chronic necro-inflammatory liver disease (Cougot *et al.*, 2008). Persistence of high HBV DNA concentration suggested an increased risk of carcinogenesis (Kumar *et al.*, 2006b).

Oncogenic RNA Viruses

The RNA viruses have also been found to cause various cancers, as discussed below:

1. Hepatitis C Virus (HCV)

HCV is an emerging infection in India which causes liver desease. It is a RNA virus, belonging to Flaviviridae family and genus *Hepacivirus* (Mukhopadhya, 2008). HCV is also strongly linked to pathogenesis of HCC. Role of this virus is related to its ability to cause chronic liver cell damage that is accompanied by liver regeneration. The mitotically active hepatocytes, surrounded by an altered environment are prone to genetic instability and cancer development (Butel, 2000). Few studies from India have also corroborated the association between HCV and HCC; the earliest report from Delhi noted that 15 per cent of patient with HCC were positive for antibody to HCV (Kumar *et al.*, 2006b).

2. Human T-Cell Leukemia Virus Type 1 (HTLV1)

Only one human retrovirus, HTLV1 is firmly implicated in the causation of cancer. HTLV1 is associated with a form of 'T' cell leukemia/lymphoma that is endemic in certain parts of Japan and the Caribbean basin, but is found sporadically elsewhere, including the United States. Similar to the AIDS virus, infection of HTLV1 in human requires transformation of infected 'T' cells via sexual intercourse, blood products or breast-feeding. Leukemia develops in only 3 to 5 per cent of the infected individuals after a long latent period of 40 to 60 years (Kumar *et al.*, 2006b).

Cancer Induced by Chemical Agents in Fish

Fish can also suffer from various types of cancer. Many cancers in fish appear to be the result of exposure to different environmental pollutants. The data correlated with contaminant or pollutant exposure and laboratory induction experiments reinforce the idea of chemical carcinogenesis in wild fish population. Liver neoplasms in sauger and walleye are associated with heavy loadings of extremely fine particulates which were produced when copper *'stamp sands'* were reprocessed. Now, one in four of the St. Lawrence whales are dying from cancer, mostly intestinal cancer. When scientists examined their bodies, the autopsies revealed high levels of PAHs. Epidemics of liver cancer have been found in 16 species of fish in 25 different polluted freshwater and saltwater locations. The same tumours were found in bottom-feeding fishes in industrialized and urbanized areas. Thus, the pollution of rivers and streams with chemical contaminants has become one of the most critical environmental problems. Fish living in a polluted water reservoir use contaminated water to rinse their gills, resulting into the deposition of PAHs in fish body. Contamination of foodstuffs by heavy metals (*e.g.*, As, Cd, Cr, Ni and Pb) has posed a potential carcinogenic threat to humans. The As and Cd appear to be the most harmful to fish. Many cancers in fish appear to be the result of exposure to various environmental chemicals. High frequencies of liver and skin cancers in brown bullheads are associated with high concentrations of PAHs and some metals in the environmental sediments (Madhuri *et al.*, 2012a; Pandey *et al.*, 2013b).

Therefore, fishes are exposed from different chemicals, including drugs and pollutants. The most important chemicals causing severe toxicity to fish are '*heavy metals*' such as As, Cd, Pb, Ni, copper (Cu), chromium (Cr), iron (Fe), manganese (Mn), mercury (Hg), zinc (Zn), tin (Sn), etc. These chemicals are also responsible to cause fish cancer. Besides these, DEN, high concentrations of PAHs and some other metals cause the cancer of liver, skin and other organs of fish (Madhuri, 2012; Madhuri *et al.*, 2012e).

Cancer Model in Fish

Fish have gained increasing attention over the past three decades as valuable models for research in environmental carcinogenesis. Various fish species have been investigated as non-mammalian vertebrate models for carcinogen testing, as surrogates for understanding mechanisms of human cancer and its prevention, as feral species indicators of ecologic contamination, as indicators of potential human exposure to carcinogens in the water column or aquatic food chain, and for application as *in situ* field monitors of integrated carcinogenic hazard in groundwater near toxic waste sites. Compared to mammals, cancer induction in fish appears to be easy. The carcinogens can be exposed through aquarium water and liver cancer induction, in particular, provides best results with most of the carcinogens so far studied (Raisuddin and Lee, 2008).

Fish has been used as a model for cancer research. One of the very popular freshwater tropical fish, *Danio rerio* (zebrafish), has emerged as a major model organism for biomedical research, especially in developmental genetics, neurophysiology, oncology and biomedicine. Like most other biomedical model organisms, zebrafish is chosen for particular traits that make it convenient for laboratory study. The greatest advantage of zebrafish as a model system comes from its well-characterized genetics, genetic and developmental techniques and tools, and availability of well-characterized mutants. Induction of experimental diseases, including cancer using zebrafish as a model organism may be a new and an important field of research (Pandey *et al.*, 2012 and 2013b).

In the 1960s to mid-1970s, the fish species used as models for carcinogenesis studies were primarily the zebrafish (*Danio rerio*) and the guppy (*Poecilia reticulata*). Species that have predominated in later years can be divided into two groups: the larger fish, *e.g.*, rainbow trout, and the small aquarium fish, including rivulus (*Rivulus marmoratus*), guppy (*P. reticulata*), sheepshead minnow (*Cyprinodon variegatus*), platyfish/swodtail (*Xiphophorus/Platypoecilus* spp.), *Fundulus grandis*, *Gambusia affinis*, *Cyprinodon variegatus* and medaka (*Oryzias latipes*). The contaminations associated neoplasia, including aflatoxin-induced HCC in rainbow trout (*Onchorynchus mykiss*) fish, have also led to the study of fish as alternative models in carcinogenesis and toxicity bioassay (Bunton, 1996; Pandey *et al.*, 2014). Teleost fish cell lines have been developed from a broad range of tissues such as ovary, fin, swim bladder, heart, spleen, liver, eye muscle, vertebrae, brain and skin. The 124 new fish cell lines from different fish species ranging from grouper to eel have been reported. Recently, about 283 cell lines have been established from finfish around the world (Lakra *et al.*, 2011; Pandey *et al.*, 2014). Further, zebrafish is increasingly recognized as a promising animal model for cancer research, and liver is the main target organ for tumorigenesis, especially HCC after carcinogen treatment. Zebrafish is a responsive, cost-effective lower vertebrate model system to study the mechanism of carcinogenesis. Liver has been found to be the most commonly targeted organ in the cancer studies with zebrafish. The neoplasms which were noticed in zebrafish include seminoma, HCC, hepatocellular adenoma (HCA), adenoma of exocrine pancreas, intestinal adenocarcinoma, ultimobranchial neoplasm, thyroid neoplasm, spindle cell sarcoma and hemangioma (Madhuri *et al.*, 2012f).

Chemical Carcinogenesis in Fish

Many studies have shown development of cancer by different chemical carcinogens in fish, as discussed below:

1. Diethylnitrosamine (DEN)

DEN (a nitrosamine or N-nitroso compound) is one of the most potent carcinogens. Primary neoplasms with histological characteristics of HCC were observed in fish exposed to 125 ppm of DEN for 3 to 5 successive periods (Pandey, 2011a). At the 18th week, hemangiomas, cholangiomas, biliary cystadenomas, and glandular, trabecular and anaplastic HCC in *R. marmoratus* fish exposed to 95 or 200 mg per L DEN for 6 weeks were observed. It was then discovered that the groundwater contaminated with trichloroethylene (TCE) has carcinogenic properties beyond what had been shown with TCE alone, suggesting that the unidentified compounds in the mixture might have promoting properties alone or synergistically with TCE. The rainbow trout were fed with a diet containing indole-3-carbinol (2000 ppm), β-naphthoflavone (500 ppm) or aroclor 1254 (100 ppm) for 6 weeks before a single 24 hr exposure of an aqueous solution of 250 ppm DEN. After 42 weeks, DEN produced 80.2 per cent incidence of liver tumours. The tumour was inhibited by indole-3-carbinol but enhanced by β-naphthoflavone. The liver cytotoxic alterations of adult medaka fish were seen following short-term bath exposure (48 hr) to 500 mg/L DEN for 3 to 21 days. Progression of hepatic neoplasia was observed in adult medaka fish (3-6 months old) following aqueous exposure to DEN (50 ppm for 5 weeks). DEN-induced spongiosis hepatis (a hepatic lesion characterized by mutilocular cyst-like complexes) was noticed in the medaka. The induction of extrahepatic tumours by DEN in fish is uncommon with the exception of pancreatic neoplasms, as seen in rivulus. DEN exposure for 2 hr was sufficient to induce the hepatic neoplasms in *K. marmoratus*. Tumour after exposure to carcinogen for such a short duration was amazing, as in other fish models it takes longer time and sometime repeated exposure is also needed. In zebrafish, about 8 weeks of exposure to DEN is needed to induce tumours. Carcino-embryonic antigen in DEN-induced liver, gut and biliary neoplasms in *K. marmoratus* fish suggested that the response was similar to mammals. Hepatomas were seen in 6- and 9- month groups of guppies (*P. reticulata*) treated with multiple doses of 100 ppm- and 200 ppm- DEN. Neoplastic foci of mixed hepatocytes and cholangiocytes increased in livers of guppies from 2nd month, developing into hepatoblastomas, which occurred in almost 100 per cent of guppies by 12th month (Pandey *et al.*, 2013b).

2. N-Methyl-N'-Nitro-N-Nitrosoguanidine (MNNG)

N-methyl-N'-nitro-N-nitrosoguanidine (MNNG) is a direct-acting carcinogen which induces neoplasms of the gastrointestinal tract (GIT) in rats, hamsters and dogs with oral administration. As a group, the N-nitroso carcinogens cause liver neoplasia in essentially all fish species, when exposure occurs in early life stages. Liver neoplasms are also reported, as well as fibrosarcoma with subcutaneous injection and multiple organ involvement with intraperitoneal injection. In addition to liver neoplasia, MNNG causes neoplasia in gill, pseudobranch, thyroid, gonad, pancreas, gas bladder, mesothelium, scale, skin, olfactory epithelium, connective tissue, skeletal muscle, notochord, blood vessels and pigment cells of medaka. In rainbow trout, only epithelial neoplasms occur in MNNG-treated fish, with liver, stomach, swim bladder and kidney targeted in trout given aqueous exposures as embryos or fry. Neoplasia occurs only in stomach in rainbow trout fed MNNG, beginning at 2 months of age. Channel catfishes given subchronic aqueous exposure to MNNG beginning at 6 months of age develop neoplasia in skin, fins, gills, thymus, bone, oropharynx and generalized lymphoid system. A high incidence of thyroid neoplasia has been reported in rivulus following immersion exposure of young fry to MNNG. MNNG also causes osteosarcoma of ribs and vertebrae in coho salmon treated as embryos. Embryos and fry were both quite responsive to MNNG; however, juvenile zebrafish were remarkably refractory to MNNG-induced neoplasia. The main target organs in zebrafish treated as embryos with MNNG were the liver and testis, with HCA as the most prevalent hepatic neoplasm. A variety of mesenchymal neoplasms occurred in zebrafish following embryo exposure to MNNG, including chondroma, hemangioma, hemangiosarcoma, leiomyosarcoma and rhabdomyo-sarcoma. The testis and blood vessels were the primary target organs for MNNG following fry exposure, with seminoma, hemangioma, hemangiosarcoma, and various other epithelial and mesenchymal neoplasms (Bunton, 1996; Pandey *et al.*, 2012, 2013b, 2013c and 2014).

With exposure of either juvenile or adult medaka, neoplasms were primarily induced in gills, connective tissues, skin, and olfactory and reproductive systems. Within the skin, neoplasms were variable and included melanoma. Vascular tumours were the most common type of neoplasm in one study, but many other tissues were affected. Multiple tissue involvement was mentioned in another juvenile exposure study, but only data on gill neoplasms were given. Despite the preceding variations, the tissue distribution of neoplasms suggests direct action of MNNG on exposed tissues. The reported distribution of neoplasms in rainbow trout exposed to MNNG was different from medaka, with no apparent relationship to exposure age or route. With embryos, fry, or adults exposed aqueously or by injection or diet, neoplasms were seen in liver, stomach and kidney, with a low incidence in swim bladder; however, in two of these reports, histologic examination was restricted to those same tissues. Liver neoplasms were relatively common in rainbow trout, whereas in medaka they were absent, not reported, or occurred at rates no higher than controls. In a separate report of 9,802 exposed (various compounds) and control medaka, 40 cases of lymphoma were seen, and the lack of a significant difference between control and treated groups indicated that the neoplasm may be spontaneous in this species. However, the tumour biology of lymphosarcoma in catfish is unknown, and an 8.8 per cent incidence of lymphosarcoma in *Poeciliopsis* strain *P. monacha* exposed to DEN was seen. Most of the neoplasms were spermatocytic seminomas containing spermatocytes and spermatids. Fewer seminomas were comprised strictly of spermatogonia, as previously reported in medaka. Interestingly, spermatocytic seminoma occurs in female medaka as well as male, in the case of both spontaneous and carcinogen-related gonadal neoplasia (Bunton, 1996; Pandey *et al.*, 2012, 2013b and 2014).

3. Benzo(a)pyrene [B(a)P]

B(a)P is an indirect-acting carcinogen. It is hepatocarcinogenic in rainbow trout, but its long-term exposure through diet or intraperitoneally is required. The racemic($\pm$)-trans-B(a)P-7,8-dihydrodiol is a much more potent carcinogen in trout. Reconstitution studies with purified enzyme and liver microsomes from BNF-treated trout indicate that CYPIA is the predominant subfamily involved in B(a)P and B(a)P-7,8-dihydrodiol bioactivation to the ultimate carcinogen 7S-trans-7,8-dihydrobenzo[a]pyrene-7,8-diol-anti-9,10-epoxide. Topical administration of a PAH extract or B(a)P also induced epidermal papillomas in two species, and a solitary gill hemangioma was seen in a high-dose B(a)P exposure group. The B(a)P is reported to cause liver cancer in guppy and madaka (Bunton, 1996; Pandey *et al.*, 2012, 2013b and 2014).

4. Dimethylbenz(a)athracene (DMBA)

DMBA is also an indirect-acting carcinogen. Although pancreatic neoplasms occurred, the incidence in DMBA- and B(a)P- exposed medaka was not higher than controls, and just one pancreatic adenoma was reported in the PAH extract exposure studies. It was suggested that the pancreatic acinar carcinoma may represent a spontaneous rather than carcinogen-induced neoplasm in medaka, but incidence of pancreatic carcinoma with MNNG exposure suggests that it may be carcinogen-dependent. A number of extrahepatic neoplasms were induced by DMBA in medaka, but the types were not specified. However, neoplasms were reportedly absent in guppy exposed to DMBA by intramuscular, intraperitoneal, or topical routes (Bunton, 1996; Pandey *et al.*, 2012, 2013b and 2014).

5. N-Methyl-N-Nitrosourea (MNU)

MNU is an alkylating agent which methylates the DNA bases primarily at nucleophilic sites (N^7 and N^3 alkylpurines). This is an indirect-acting carcinogen. The primary mutagenic lesion of MNU exposure is believed to be O^6-methylguanine. The MNU induces several cancers in rodents, including mammary carcinomas and thyroid tumours in rats. It has also induced various tumours in *Xiphophorus* hybrids, including neuroblastomas, melanomas, fibrosarcomas, rhabdomyosarcomas at high incidence, and various carcinomas at a greatly reduced incidence (Bunton, 1996; Pandey *et al.*, 2012, 2013b and 2014).

6. Other Chemicals

Cancers have been reported in brown bullheads in association with PAH and other contaminants. Other similarly affected benthic species include black bullheads (*Ictalurus melas*) exposed to chlorinated wastewater effluent, white suckers (*Catostomus commersoni*) and white croakers. Increased incidence of papillomas is most commonly seen in polluted sites, papillomas have been seen in brown bullhead populations. Besides, certain other chemicals are also responsible for carcinogenesis in fishes, as mentioned in Table 1 (Bunton, 1996; Pandey *et al.*, 2013b and 2014).

Table 1: Certain Chemical Carcinogens Causing different Cancers in Fish

Carcinogen	Fish Species	Type of Cancer
Aflatoxin B_1 (AF B_1)	*Poecilia reticulata* (Guppy)	Hepatic cancer
	Oryzias latipes (Madaka)	Hepatocellular carcinoma (HCC), hepatic adenoma (HA)
	Oncorhynchus mykiss (Rainbow trout)	HCC
N-nitrosomorpholine(NM)	*Poecilia reticulata*	HCC, cholangiocellular carcinoma (CCC), oesophageal cancer
	Danio rerio (Zebrafish)	HCC, CCC, oesophageal cancer
Orthoaminoazotoluene(o-AAT)	*Poecilia reticulata*	HCC, CCC
	Oryzias latipes	HCC, HA
N-nitrosodimethylamine (DMN)	*Danio rerio* (Zebrafish)	HCC, CCC, oesophageal cancer
	Poecilia reticulata	HCC, CCC
2-AAF	*Lebistes reticulates* (Guppy)	Hepatic cancer
Methylazoxymethanol acetate (MAMA)	*Gambusia affinis*	HCC, CCC
	Oryzias latipes	HCC, medullo-epithelioma
	Poecilia reticulata	Pancreatic acinar cell carcinoma (ACC), adenocarcinoma (AC)
Butylated hydroxyanisole (BHA)	*Rivulus ocellatus* (*marmoratus*)	HCC
Dichlorodiphenyl-trichloroethane (DDT)	*Oncorhynchus mykiss*	HCC

Cancer by Pollutants in Fishes

The term *'pollution'* covers all types of contamination emitted to the ground, surface water and atmosphere that could be further absorbed by plants, fish and farm animals. Sources of pollution include industrial waste, diesel exhaust (mostly in the vicinity of roads) and pesticide residues in food products. An example of the effect of environmental pollution on nutrition is contamination of fish by PAHs. Another example is contamination of foodstuffs by heavy metals. Although some metals are essential for human nutrition, others including As, Cd, Cr, Ni and Pb have also posed a potential carcinogenic threat to humans. The As and Cd are the most toxic to fish (Madhuri *et al.,* 2012a). Thus, the pollutants can cause chemical carcinogenesis in fish (Madhuri, 2012).

In fact, the contamination of freshwater with chemicals or pollutants has become a matter of concern over the last few decades. The aquatic systems may extensively be contaminated with heavy metals released from domestic, industrial and other man-made activities. Heavy metal contamination may have devastating effects on ecological balance of the recipient environment and a diversity of aquatic organisms. Because of the pollutants, a huge mortality occurs in fish. Different types of toxicity, leading to various diseases and cancer caused by heavy metals, *e.g.*, As, Cd, Pb and Hg is most common in fish. In cooperation with the US Environmental Protection Agency, the *'Agency for Toxic Substances and Disease Registry'* in Atlanta, Georgia (a part of the US Department of Health and Human Services) reported that in a *'Priority List for 2001'* called the *'Top 20 Hazardous Substances'*, As, Pb and Hg are at the 1st, 2nd and 3rd position, respectively in the list; while Cd is at the 7th place. The As is a common element that occurs in air, water, soil and all living tissues. It ranks 20th in abundance in the earth's crust, 14th in seawater and 12th in the human body. It is a carcinogen and can cause foetal death and malformations in many species of mammals (Madhuri *et al.,* 2012e).

The distribution of heavy metals varies between fish species, depending on age, development status and other physiological factors. Fish accumulate substantial concentrations of Hg in their tissues and thus can represent a major dietary source of this element for humans. Fish are the single largest sources of As and Hg for humans. The primary sources of Hg contamination in humans are through eating fish. Metal contaminations in food, especially in marine products, have been broadly investigated (Emami Khansari *et al.,* 2005). Among animal species, fishes are the inhabitants that can not escape from the detrimental effects of these pollutants. The studies carried out on various fishes have shown that heavy metals may alter the physiological activities and biochemical parameters both in tissues and in blood (Vinodhini and Narayanan, 2008). The liver, kidney and gill of fish are the main organs for pollution. Liver and gills as main organs for metabolism and respiration are target organs for contaminants accumulation, and after the exposure of pollutants, the damage to organs and tissues occurs (Madhuri *et al.,* 2012e; Montaser *et al.,* 2010).

Because of toxicity and accumulative behaviour of heavy metals, they can make various changes in aquatic environment like species diversity. Heavy metals can enter into water via drainage, atmosphere, soil erosion and all human activities by many ways. With increasing heavy metals in the environment, these elements enter the biogeochemical cycle. They can enter from contaminated water into fish body by different routes and accumulate in various organs. Some heavy metals, *e.g.*, Fe, Cu, Zn and Mn are essential for biological systems like enzymatic activities; whereas, other heavy metals like Pb, Cd and Hg have no known important role in living organs and are toxic even in trace amounts. Essential metals must be taken up from water, food or sediment by fish for its normal metabolism; however, these metals can also have toxic effects at high concentration (Dobaradaran *et al.,* 2010). The metal pollution may damage marine organisms at the cellular level and possibly affect the ecological balance. Exposure and ingestion of polluted marine organisms as sea foods can cause health problems in people and animals, including neurological and reproductive problems. Chemicals of industrial effluents and products of ships and boats, which find their way into different water systems, can cause toxic effects in aquatic organisms. Petroleum products are one of the most relevant pollutants to aquatic ecotoxicology. Exposure to crude oil and derivatives can induce toxic symptoms in experimental animals. Petroleum hydrocarbons can act as a mediator in free radical generation in fish. The coastal pollution occurs with several pollutants, including heavy metals (Montaser *et al.*, 2010).

Indeed, it is essential to know how to examine the fish for cancers, to recognize gross signs of tumours and know how to prepare samples for histopathology. Therefore, the field biologist should develop skill in recognizing the presence of commonly occurring cancers in fish and in preparing samples for further diagnosis. Zebrafish are used to study the development, toxicology and toxicopathology. The zebrafish is known for its rapid development in the laboratory. Expansion of synthetic chemical producing industries during the 1940's coincided with a number of pollution-associated fish neoplasia epizootics, with PAHs as significant components of contaminated sediments in several cases. Epizootics of primarily liver and skin neoplasia in benthic species near coastal urban or industrial areas indicated the sensitivity of fish species to known mammalian carcinogens. The potential for application of research findings to both human and environmental health issues make fish species attractive and valuable alternative models in the carcinogenesis and toxicity research (Pandey *et al.*, 2012).

Certain Pollutants Causing Fish Cancer

A direct-acting carcinogen, MNNG induces many cancers, as stated earlier in this chapter (Pandey *et al.*, 2013b). Carcinogenic effects of certain N-nitroso carcinogens have been studied in some species. A high incidence of thyroid neoplasia is reported in rivulus, following immersion exposure of young fry to MNNG. The MNNG also causes osteosarcoma of ribs and vertebrae in coho salmon treated as embryos. Benign or malignant thyroid neoplasia is the most common histologic type of epithelial neoplasia reported in medaka following fry treatment with MNNG; and seminoma is second most frequent histologic type of epithelial neoplasia seen in medaka given fry bath treatment with MNNG. The MNNG and other N-nitroso carcinogens induce pigment cell neoplasia in medaka, Nibe croaker and *Xiphophorus* (Bunton, 1996; Pandey *et al.*, 2012).

Another pollutant is the DEN, also called N-nitrosodiethylamine (DENA). It is one of the most potent carcinogens. Much of its carcinogenic effects have been discussed earlier in this chapter (Pandey *et al.*, 2013b). In the 1970s, there was an increased frequency of liver cancer found in Norwegian farm animals, which were fed on herring meal, which was preserved using sodium nitrite. The sodium nitrite had reacted with dimethylamine in the fish and produced DEN. Nitrosamines can cause cancers in a wide variety of animal species, a feature that suggests that they may also be carcinogenic in humans. DEN has been extensively used in medaka and rainbow trout, and has proved to be a potent inducer of hepatic neoplasms (Bunton, 1996; Pandey *et al.,* 2012).

MAMA is also a potent carcinogen with direct- and indirect- acting properties. It induces liver, kidney and colon cancers in rodents and nonhuman primates when administered by different routes. B(a)P is an indirect-acting carcinogen, as described earlier in this chapter (Pandey *et al.,* 2013b). It causes hepatocarcinogenesis in rainbow trout but its long-term exposure through the diet or intraperitoneally is needed (Bunton, 1996; Pandey *et al.,* 2012 and 2014). DMBA is another indirect-acting carcinogen, discussed earlier in this chapter (Pandey *et al.,* 2013b). This pollutant chemical compound is much more potent hepatocarcinogen in trout than B(a)P and produces tumours in kidney, swim bladder and stomach as well. Preliminary evidence indicates that DBP resembles DMBA with respect to trout target tissues but is a more potent carcinogen, especially for liver and swim bladder. It has been reported that DMBA causes HCC, hepatic adenoma, renal acinar cell carcinoma, neurilemmoma and fibrosarcoma in guppy. It has also caused HCC and lymphosarcoma in *Poeciliopsis* species. In *Cyprinodon variegates* also, DMBA caused HCC and hepatic adenoma (Bunton, 1996; Pandey *et al.,* 2012 and 2014). MNU is an indirect-acting carcinogen as described earlier in this chapter (Bunton, 1996; Pandey *et al.,* 2012 and 2013b). The papillary thyroid tumour induction by MNU has also been reported (Lee *et al.,* 2000).

Liver and Skin Cancers Caused by Chemical Contaminants in Fish

The most important chemical contaminants causing severe toxicity to fish are *'heavy metals',* e.g., As, Cd, Cr, Ni, Pb, Cu, Fe, Mn, Hg, Zn and Sn, etc. Similarly, the high concentrations of PAHs and some metals in environmental sediments cause liver and skin cancers in brown bullhead fishes. Thus, the chemical contaminants are responsible for induction of carcinogenesis in fishes. The fishes living in a polluted water reservoir (in the vicinity of an oil refinery) use the contaminated water to rinse their gills, resulting into the deposition of PAHs in the fish body. The contamination of foodstuffs is also by the heavy metals. Although some heavy metals are essential for human nutrition, others including As, Cd, Cr, Ni and Pb, have been found to pose a potential carcinogenic threat to humans as well as fish, and Cd appear to be the most harmful. In populations near urban centres where concentrations of PAHs, PCBs, DDT and other compounds were elevated in tissue (stomach and liver) and sediment, there was an increased chemical-associated risk of developing cancers and associated lesions (Pandey *et al.,* 2013c).

Scientific and public interest in the occurrence of tumours (neoplasms) began slowly in the 1960s; and by 1986, systematic tumour surveys started in some parts of the world. Contaminant surveillance programmes have reported that the environment is receiving and accumulating an array of organic and inorganic contaminants. The presence of carcinogens, combined with fish tumour induction data and reports of tumours in wild fish populations, have provided strong circumstantial evidence that chemical carcinogenesis is occurring at some sites (Pandey *et al.,* 2013c). Liver has been found to be the most targeted organ for fish cancer. The cancers which have been reported in fish (particularly in zebrafish) are seminoma, HCC, HCA, adenoma of exocrine pancreas, intestinal adenocarcinoma, ultimobranchial neoplasm, thyroid neoplasm, spindle cell sarcoma and hemangioma, etc. (Madhuri *et al.*, 2012f).

Liver Cancers in Fish by Chemical Contaminants

High prevalence of liver (hepatic) cancers and related lesions have been reported in mummichog (*Fundulus heteroclitus*) fish from a PAH-contaminated site in the Elizabeth river along the mid-Atlantic coast. The PAH sediment concentrations were associated with active and abandoned wood creosote treatment facilities near the site, which experienced several significant creosote spills in the 1960s. Using hepatoproliferative lesions in rodents as a guide, the lesions included eosinophilic, basophilic and clear cell foci of cellular alteration, HCA and HCC, and proliferative biliary lesions, which were not seen in populations from less contaminated sites. The epizootic hepatic neoplasia was also reported from the North Atlantic coast in winter flounder (*Pleuronectes americanus*), a valuable, benthic (bottom dwelling) fish. The contaminated sites were associated with outfall from sewage treatment plants and dredge spoils. The dredge sediments contained a complex mixture of PAHs, PCBs, metals and other compounds. The hepatic lesions included basophilic foci of cellular alteration, HCA, HCC, cholangiocarcinoma and proliferative hydropically vacuolated ductular epithelial cells. The epizootic of hepatic neoplasia in another benthic fish, the English sole (*Parophrys vetulus*), on the opposite coast of the United States, have been recorded. In Puget Sound of Washington State, the cancers and associated lesions were seen only in fish from estuaries and embayments proximal to urban centers or pollution point sources. The sediments contained a wide range of chemicals, which included B(a)P and benzo(a)anthracene, heavy metals, PCBs and other chlorinated hydrocarbons. The lesions in juveniles included degenerative megalocytic hepatosis frequently accompanied by eosinophilic, basophilic and clear cell foci of cellular alteration. Benign and malignant hepatocellular and biliary cancers were seen in adult fish of at least two years of age. The English sole injected intramuscularly with B(a)P or a PAH-enriched fraction of sediment extract from Eagle Harbor also resulted in megalocytic hepatosis and foci of cellular alteration, providing a link between exposure and development of these lesions (Buntun, 1996; Pandey *et al.*, 2013c).

Different carcinogens or chemicals have been reported to produce the experimental liver cancers in fish. Primary neoplasms with histological characteristics of HCC were observed in fish exposed to 125 ppm of DEN for 3 to 5 successive periods. In another study, hemangiomas, cholangiomas, biliary cystadenomas, and glandular, trabecular and anaplastic HCC were observed at the 18th week in rivulus fish exposed to 95 or 200 mg per L DEN for six weeks (Bunton, 1996). The HCC was induced by DEN in medaka fish as early as five weeks post-exposure. The rainbow trout were fed with a diet containing indole-3-carbinol (2000 ppm), β-naphthoflavone (500 ppm) or aroclor 1254 (100 ppm) for six weeks before a single 24 hr exposure of an aqueous solution of 250 ppm DEN. After 42 weeks, DEN produced 80.2 per cent incidence of liver tumours. Liver cytotoxic alterations of adult medaka (*Oryzias latipes*) fish were seen following short-term bath exposure (48 hr) to 500 mg per L DEN for 3 to 21 days (Bunton, 1996; Pandey, 2011a; Pandey *et al.*, 2013c). Progression of hepatic neoplasia was observed in adult medaka fish (3-6 months old) following aqueous exposure to DEN (50 ppm for 5 weeks) (Okihiro and Hinton, 1999). The DEN-induced spongiosis hepatis (a hepatic lesion characterized by mutilocular cyst-like complexes) was noticed in the Japanese medaka, *O. latipes* (a small aquarium fish) (Norton and Gardner, 2005). Induction of extrahepatic tumours by DEN in fish is uncommon with the exception of pancreatic neoplasms, as seen in rivulus fish. DEN exposure for 2 hr was sufficient to induce hepatic cancers in *Kryptolebias marmoratus*. *K. marmoratus* was used for study of oncogenes in biliary and hepatic cancers, and necrotic and regenerative phases of DEN-toxicity. Hepatomas were noticed in DEN-exposed *K. marmoratus* and *P. reticulata* (guppy) fishes (Bunton, 1996; Pandey, 2011a; Pandey *et al.*, 2012 and 2013b).

The mechanisms of hepatocarcinogenesis in fish have also been studied. As the preneoplastic hepatic lesions in fish are in many ways similar to those noticed in rodent hepatocarcinogenesis, the terminology is the same. However, the relative significance of each lesion may vary in fish cancers. A number of nonspecific changes preceding the development of preneoplastic and neoplastic lesions are important in the pathogenesis of hepatic cancer in fish. In medaka, sheepshead minnow and rivulus fishes exposed to DEN by the aqueous route, there is an early cytotoxic phase with necrosis and dropout of hepatocytes resulting in cystic degeneration. Vacuolar degeneration, fatty changes and hyalinized eosinophilic cytoplasmic inclusions may also develop and persist. The eosinophilic cytoplasmic inclusions described in DEN-exposed medaka, zebra danio, rivulus and MAMA-exposed gambusia most likely represent apoptotic bodies. The severity of necrosis appears to increase with chemical concentration but also occurs at lower exposure levels. The cytotoxic phase is followed by proliferation of epithelial and interstitial cells (Bunton, 1996).

The foci (basophilic, eosinophilic and clear cell types) of hepatocellular alteration morphologically similar to those described in rodent hepatocarcinogenesis have been reported in several fish species. All these three types of foci have been identified in hepatic neoplasia epizootics (Myers *et al.*, 1994). The basophilic foci, consisting of small RNA-rich cells as classically described in rodents, are considered to progress to HCC in rainbow trout, medaka, guppy and sheepshead minnow. The eosinophilic foci, consisting of cells containing abundant smooth endoplasmic reticulum in rodents, have also been reported in several fish species. However, the differences in the normal tinctorial properties of male and female hepatocytes related to vitellogenesis can make differentiation of basophilic from eosinophilic foci difficult in medaka. The clear cell foci are most variable of foci of cellular alteration in fish species. The foci of cells containing abundant glycogen have been reported in rivulus, mummichog and sheepshead minnow but less commonly in medaka and rainbow trout. The foci of vacuolated cells have been described as the most likely contained lipid in medaka, and a mixture of lipid and glycogen in rivulus. Proliferation of bile ductular epithelial cells and cholangiofibrosis are common findings, and compatible non-neoplastic counterparts of the commonly seen benign and malignant biliary neoplasms. Oval cells, a proliferative population of ductular epithelial cells common in rodent hepatocarcinogenesis are seen in the sheepshead minnow and medaka exposed to DEN, and these have been postulated as the cell type responsible for regeneration after cytotoxic dosages of aflatoxin B_1 in rainbow trout (Bunton, 1996). A distinct lesion of hydropic vacuolization of proliferative preductular cells seen in winter flounder and other benthic species is considered to be a biomarker of environmental chemical contamination. Correlation of this lesion with rodent oval cell hyperplasia has been suggested (Moore and Stegeman, 1994).

The proliferation of hepatic interstitial cells in response to carcinogenic exposure is an important component of fish carcinogenesis, perhaps more so than in rodents. Two lesions are of particular importance. *'Spongiosis hepatis'*, a multilocular pseudocystic lesion induced by N-nitroso compounds in rats, commonly occurs in fish hepatocarcinogenesis bioassays with a variety of compounds. It appears most commonly in the small aquarium species but is also seen rarely in benthic species from hepatic cancer epizootics (Myers *et al.*, 1994). The lesions first appear after the cytotoxic phase, with subdivision of cystic spaces by perisinusoidal cells, and can become extremely enlarged or aggressively multifocal. In both rats and fishes, the spongiosis hepatis may be associated with foci of cellular alteration or hepatocellular cancers but in fishes, the association appears to be random. Another interstitial cell lesion, *'spindle cell proliferation'* consists of an unspecified population of spindle-shaped cells which replaces portions of the hepatic parenchyma. Although spindloid biliary epithelial cells have been found to comprise portions of some lesions by keratin immunocytochemistry, a substantial population remains uncharacterized and may include fibroblasts, endothelial cells and/or perisinusoidal cells. Fibroplasia with deposition of collagen is uncommon in fish studies but a mesenchymal reaction, including cirrhosis has been noticed in medaka exposed to MAMA and other compounds. In medaka, the perisinusoidal cells have been shown to express actin in trabecular and schirrous HCC induced by DEN or MAMA. This change, also seen in mammals, is interpreted to be a phenotypic reaction to injury.

However, perisinusoidal cells, spongiosis hepatis lesions and hyperplastic perisinusoidal cell aggregates in rats are also desmin positive, whereas comparable lesions in fish have been found negative, and the decreased desmin antibody sensitivity in fish tissues may have affected the results. The studies of DEN-induced toxic lesions and hemangiopericytoma have identified proliferative cells with features shared by both perisinusoidal cells and endothelial cells, raising the possibility of a common stem cell between these cell types. The most common type of interstitial cell neoplasm reported to occur in several species is hemangiopericytoma (Bunton, 1996; Pandey *et al.*, 2013c).

Skin Cancers in Fish by Chemical Contaminants

Epizootics of skin cancer/neoplasia have also been associated with chemical contamination. In 1941, Lucke and Schlumberger reported an epizootic of orocutaneous papillomas in brown bullhead (*Ictalurus nebulosus*) in the Delaware and Schuylkill rivers near Philadelphia. Since then, liver and skin cancers have been reported in brown bullheads in association with PAHs and other contaminants from the Great Lakes to New York State and Ontario (Canada). Other similarly affected benthic species include black bullheads (*Ictalurus melas*) exposed to chlorinated wastewater effluent, white suckers and white croakers (Bunton, 1996). Although increased incidence of papillomas is most commonly seen in polluted sites, the papillomas have been seen in brown bullheads from relatively clean sites (Poulet *et al.*, 1994). In addition, lesions on white suckers observed in the laboratory regressed under uncrowded conditions (Premdas and Metcalfe, 1994).

However, the papillomas have been induced in brown bullheads and mice by epidermal applications of heavily polluted sediment extract containing PAHs (Bunton, 1996). The epidermal neoplasms in brown bullheads appear to progress through the classic stages of skin carcinogenesis, including hyperplasia, papilloma and carcinoma (Poulet *et al.*, 1994), although carcinoma is less commonly seen. The endophytic hyperplasia appeared to progress to large fungating exophytic papillomas. Focal invasion through the basal lamina marked the transition from papilloma to carcinoma, similar to other report (Bunton, 1996; Pandey *et al.*, 2013c; Poulet *et al.*, 1994).

Estrogen Causes Cytotoxicity and Cancer

Estrogens: Their Sources and Uses

'*Estrogens*' (or oestrogens) are the female hormones, mainly secreted by ovary and placenta; small amounts may be produced by the adrenal glands and testes in humans. In certain animals like horse, large quantities of estrogens are produced by the testes. The estrogens are also present in various plants and seeds. The natural estrogens are steroids. However, typical estrogenic activity is also shown by chemicals which are not steroids. Thus the term '*estrogen*' is used as a generic term to describe all the compounds having an estrogenic activity (Madhuri, 2008; Madhuri *et al.*, 2012d; Satoskar *et al.*, 2005).

Estrogens have been classified into three groups (Loose and Stancel, 2006; Satoskar *et al.*, 2005): (a) natural estrogens, *e.g.* (o)estradiol, (o)estrone and (o)estriol, and their esters; (b) semisynthetic estrogens, *e.g.*, ethinyl (o)estradiol (EO), mestranol and tibolone; and (c) synthetic estrogens- steroids, *e.g.*, mestranol; and nonsteroids, *e.g.*, stilbene, or derivatives of diphenylethylene (*viz.*, stilbestrol or DES) and triphenylethylene (*viz.*, chlorotrianesene). The DES and chlorotrianisene are now used only to treat patients with carcinoma of the prostate and breast. Other nonsteroidal estrogens include hex(o)estrol, dien(o)estrol and quin(o)estrol, etc.

Pituitary gonadotropins are important in regulating the production of ovarian steroids. '*Luteinizing hormone*' (LH) stimulates the production of two androgens, *viz.*, androstenedione and testosterone by '*theca cells*' of ovarian follicles. These androgens are taken up by the adjacent '*granulosa cells*' of ovarian follicles. The granulosa cells proliferate and express aromatase under the influence of '*follicle stimulating hormone*' (FSH). Aromatase converts androstenedione and testosterone into estrogens, estrone and estradiol, respectively. The estrogens are also synthesized in the testes and placenta. The hepatic and adipose tissues contain aromatase, and may convert the circulating androgens into estrogens. At puberty, the output of gonadotropins increases with corresponding activation of the gonads, leading to an increased production of ovarian estrogens. In female, the estrogen stimulates growth and development of uterus at puberty, causes endometrium to thicken during first half of the menstrual cycle and influences breast tissues throughout the life, but particularly from puberty to menopause. Of the three natural estrogens, estradiol ($17b$-estradiol) is the most potent and major secretory product of the ovary. It is readily oxidized to estrone, which in turn can be hydrated to estriol. These transformations take place mainly in the liver, where there is free interconversion between estrone and estradiol. All these three estrogens are excreted in the urine as glucuronides and sulphates, along with host related, minor products in water soluble complexes (Loose and Stancel, 2006; Madhuri *et al.*, 2012d; Satoskar *et al.*, 2005).

Some of the herbs with estrogenic activity include anise, hops, fennel, milk thistle, red clover, licorice, royal jelly, sage, fenugreek, burdock and rhubarb. However, some of the foods with estrogenic activity are French bean, date palm, dates, garlic, apple, soy(a)bean, soya sprouts, chick pea, cherry, alfalfa, cow pea, green bean, red bean, raspberry, carrot and squash (Thompson *et al.*, 2006). Some chemical compounds which have been reported (Mukherjee *et al.*, 2006) to possess estrogenic activities are DDT, kepone, methoxychlor and toxaphene (insecticides), atrazine (herbicide), bisphenol A, butylhydroxyanisole and nonylphenol (plastic manufactures), dieldrin and endosulphan (pesticides), 3-(4-methylbenzylidene)-camphor (organic UV sun screen), PCBs (plasticizers, dyes and coolants), o-phenyl phenol (fungicide, disinfectant, dye and rubber), phthalates (plastics and fixatives), and DES.

As mentioned earlier, estrogens are most commonly used as OCs in premenopausal women and as HRT in postmenopausal women. They are also indicated in females for the treatment of amenorrh(o)ea, genital disorders (*e.g.*, vulvo-vaginitis in children) and acute severe dysfunctional uterine bleeding, and in male to treat prostate cancer (Loose and Stancel, 2006; Satoskar *et al.*, 2005). In bitches, the estrogens and synthetic estrogenic drugs are used for the treatment of misalliance, hypogonadal obesity and hormonal urinary inconsistence. In male dogs, the oestrogenic preparations are used to treat anal adenoma, excess libido and prostatic hyperplasia (Acke *et al.*, 2003; Cain, 2001).

In spite of its great importance and usefulness, estrogen is responsible for several detrimental effects, including cancer/carcinogenesis/carcinogenicity both in humans as well as animals. Man's great reproductive potential and the greatly increased survival rates in recent years have posed a major problem of population. Use of OCs to control the human population is one aspect. The OC drugs are tremendously used by millions of women throughout the world to prevent fertilization or to control birth. They first became available to American women in the early 1960s, since then they have influenced the lives of untold millions of women and have had a revolutionary impact on global society. The OCs contain two hormones, *viz.*, estrogen and progesterone (progestin) (Loose and Stancel, 2006; Madhuri, 2008; Madhuri and Pandey, 2010b; Madhuri *et al.*, 2007b and 2012d; Pandey and Madhuri, 2008a; Satoskar *et al.*, 2005).

Estrogen is not without Danger: It May Cause Cancer

It is disconcerting to think that a natural hormone (estrogen) circulating in significant amounts through the bodies of half of the world's population (women) is a carcinogen, but it is now official. In December 2000, The USA Government's National Toxicology Programme and the National Institute of Environmental Health Sciences have added estrogen to the list of known human carcinogens (Madhuri, 2008; Madhuri and Pandey, 2010b; NTP/NIEHS, 2001; Platt, 2005).

Estrogen hormone is headed for trouble and is responsible for many illnesses, including cancer in women as well as men. Excessive and prolonged use of estrogen has been reported to cause cancers of endometrium, cervix, vagina, ovary, liver and mammary gland in pre- or post- menopausal women (Cutler *et al.*, 1972; Edmondson *et al.*, 1976; Gambrell *et al.*, 1983; Loose and Stancel, 2006; Madhuri, 2008; Madhuri *et al.*, 2007b; Madhuri and Pandey, 2010a and 2010b; Meissner *et al.*, 1957; Satoskar *et al.*, 2005; Shar and Kew, 1982; Vessey *et al.*, 1983). Therefore, all oral forms of estrogen must be processed by the liver, which traps over 90 per cent of the hormone, allowing less than 10 per cent only to reach into the bloodstream. Indeed, the excessive estrogen is trapped in the uterus, ovary, breast, liver, or in other target organs due to stagnation, which overstimulates the cell division leading to many disorders like fibroids, cysts or cancer (Hertz, 1976; Madhuri *et al.*, 2007c; Pandey and Madhuri, 2008a). Over doses of estrogen may cause nausea, vomiting, anorexia (loss of appetite), migraine, blurring of vision, mental depression, headache, asthma, endometriosis, fibroids, breast engorgement (fullness and tenderness), increased vaginal secretion (leucorrhoea), edema/oedema, cardiovascular and hepatic diseases, cancer, stroke, Alzheimer's disease and many others in human beings (Loose and Stancel, 2006; Madhuri, 2008; Madhuri *et al.*, 2012d; Pandey and Madhuri, 2008a; Satoskar *et al.*, 2005).

Henceforth, the excessive estrogen and its stagnation in blood circulation cause cancers of many organs, including breast, uterus and ovary. Estrogen is believed to cause cancer by helping cells proliferate. After the hormone binds to its receptors in a cell, it turns on hormone-responsive genes that promote DNA synthesis and cell proliferation. In metabolizing carcinogenic estrogen, the reactions produce intermediates capable of producing 'oxygen radicals' that can damage the cell's fats, proteins and DNA. Unrepaired DNA damage can turn into a mutation, leading to cancer (Platt, 2005; Madhuri, 2008; Madhuri and Pandey, 2010b; Madhuri *et al.*, 2009).

After chronic dosing of stilbestrol, the degeneration with marked glandular atrophy was noticed in the endometrium of bitches. The animals receiving contraceptives for short periods had a thickened endometrium with numerous glands; and after a long period of contraceptive therapy atrophic endometria or squamous metaplasia of endometrium with some cases of cystic dilatation of the glands were seen. In another study, the old bitches dosed with quinestrol (an estrogen) for one year. The bitches developed vaginal exudates, swollen mammary glands, enlarged nipples, vulval swelling, alopecia, uterine enlargement with endometrial hyperplasia and myometrial hypertrophy, especially at higher doses of quinestrol. A synthetic estrogen, hexestrol (60 mg/kg, orally daily up to 40 days) was found to be hepatotoxic in female rats. In this study, there was found no mortality, and only the modest external signs of toxicity were observed. All the rats showed reduction in body weight and appetite, and gain in liver weight. This latter effect was due to both cellular hypertrophy and hyperplasia. The fatty changes in the liver were mainly noticed which affected scattered individual cells, chiefly in the midzonal region. Estradiol has been reported to cause tumors (tumours) arising from the interstitial cell hyperplasia, with proliferation and invasion of the ovarian surface epithelium into the ovarian stroma. A bitch was observed with clinical hyperoestrogenism that developed follicular ovarian cysts, and cystic endometrial hyperplasia and pyometra in the uterus (Madhuri *et al.*, 2012d).

After continuous exposure of estradiol, the formation of a papillary ovarian surface resembling human serous neoplasms of low malignant potential in laboratory animals (Bai *et al.*, 2000), and the ovarian surface epithelial cell proliferation in sheep (Murdoch and Van Kirk, 2002) were seen. EO and some other estrogens (*viz.*, 2- and 4-hydroxyestradiol, DES and estradiol) caused uterine adenocarcinoma after 12- and 18- months in mice (Newbold and Liehr, 2000). Adverse effects like bone marrow suppression, pyometra and infertility were reported (Cain, 2001) in bitches after treatment with many estrogen preparations. Further, estrogen caused lethargy, anorexia and collapse with clinical findings, including vaginal discharge, endometritis and enlarged mammary glands in bitches (Acke *et al.*, 2003). The estrogen induced endometrial hyperplasia and carcinoma in laboratory animals have also been reported (Kumar *et al.*, 2006b).

Most of the OCs contain 0.02 to 0.1 mg (20-100 µg) of EO. As little as 0.02 mg of EO, daily serves as HRT in the menopause; while 0.015 mg, daily can prevent vasomotor symptoms in menopause; and doses of 0.015 to 0.25 mg, daily are required to prevent the bone loss in postmenopausal women. For the treatment of prostate cancer, EO is administered at the dose of 0.5 to 3 mg, orally, daily (Loose and Stancel, 2006; Satoskar *et al.*, 2005); while EO (in the form of lynoral tablet) is prescribed as oestrogen replacement therapy (ORT) at the dose of 0.05 mg 1 to 3 times, orally, daily. The LD_{50} of EO has been reported to be more than 1000 µg per kg body weight, orally in rats (Madhuri *et al.*, 2007a and 2007b; Pandey and Madhuri, 2008a).

Experimental toxicity study of EO (@ 300-1000 µg/kg, orally) in female albino rats showed some gross effects like dullness, listlessness, deep breathing, decrease in spontaneous motor activity (SMA) and respiration rate, incoordination in gait and posture, tremor, piloerection, less response to stimuli, increased secretion from nose and vagina, lachrymation, redness of eye, nose and ear, breast engorgement, and loss of appetite. The toxicity of EO was found to be dose and time dependent, suggesting that the EO at higher dose for prolonged period causes more toxicity, and may also change the normal anatomy and histology of the body organs (Madhuri *et al.*, 2007c; Pandey and Madhuri, 2008a). The EO (@ 250, 500 and 750 µg/kg, orally, weekly) administered for 8 weeks to albino rats caused mild to severe congestion along with the infiltration of lymphocytes in the interstitial tissue of ovary as observed on the 9th week. However, the EO (@ 250, 500 and 750 µg/kg, orally, weekly) administered for 12 weeks produced more severe changes with marked fibrovascular connective tissue replacing the ovarian parenchyma as observed on the 13th week. At this period in higher dose (750 µg/kg), swelling of endothelial cells, thickening of blood vessel walls, degeneration and fibrous tissue proliferation were also observed (Madhuri *et al.*, 2007c). At the same doses and periods, the EO also damaged the uterine tissues of albino rats. On the 9th week, the vacuolar degeneration of columnar epithelial cells in the mucosa and endometrial glands was observed. The blood vessels in the endometrium (uterus) were dilated and congested. Degeneration and necrosis of the mucosa and endometrial gland with infiltration of inflammatory cells were also noticed. At some places, proliferation of fibrovascular connective tissues was distinct. On the 13th week, these changes were more marked (Madhuri *et al.,* 2009).

EO (250, 500 and 750 µg/kg, orally, weekly for 12, 16 and 20 weeks) administered to different groups of female albino rats injured the hepatic tissues. On the 13th week, the hepatic tissues of rats administered with EO at the dose of 250 µg per kg, showed congestion and perilobular fibrosis. However, the hepatic tissues of rats administered with EO at the dose of 500 µg per kg, revealed severe congestion, focal areas of haemorrhage, varying degree of degeneration and necrosis; central veins were extremely dilated and few of them were congested; and fibroblastic proliferation was distinct with distended sinusoids at certain places. The hepatic tissues of rats administered with EO at the dose of 750 µg per kg, showed similar changes; however, the fibroblastic reaction was more intense at the portal triad area, and the formation of new bile duct was evident. On the 17th and 21st weeks, the hepatic tissues of rats administered with EO in doses of 250 to 750 µg per kg, showed more severe and extensive changes (Pandey *et al.,* 2009 and 2011).

Furthermore, the EO induced gynaecologic (uterine and ovarian) cancer in rats have been observed (Madhuri, 2008). An experimental model of cancer in the ovary was assessed after administration of EO. This drug (750 µg/kg, orally, weekly) was administered for 16, 20 and 24 weeks in three groups of female albino rats, respectively. On the 17th week, the ovarian tissues revealed degeneration and necrosis of follicular tissues, thick banding leading to fibrosis, thickening of vascular wall and presence of homogeneous mass in the lumen. On the 21st week, fibroplasia of interfollicular connective tissues, hyperplasia of follicular cells and papillary proliferation in surface epithelium were observed. On the 25th week, besides these changes, other malignant lesions like hyperchromatosis, enlargement of nuclei and anisokaryosis, anisocytosis were seen. These changes indicated that EO caused the adenocarcinoma in the rat ovary (Madhuri and Pandey, 2011; Madhuri *et al.,* 2011a, 2012b and 2012d).

Uterine and Ovarian Cancer Models Produced by Oestrogen in Rat

Oestrogen as OC and HRT is the most commonly used by millions of women all over the world. Unfortunately, its excessive and prolonged use has been reported to cause cancers of many organs, including gynecologic cancer (Loose and Stancel, 2006; Madhuri and Pandey, 2010b; Madhuri *et al.,* 2007a and 2007b; Satoskar *et al.,* 2005).

Both exogenous and endogenous oestrogens and their metabolites play a significant role in the pathogenesis of cancer in the oestrogen-responsive tissues of humans and animals. Hence, the long-term administration of oestrogen is followed by the development of tumours of breast, cervix, endometrium, ovary, pituitary, testicle, kidney and bone marrow in either mice, rats, rabbits, hamsters or dogs (Hertz, 1976). Some other authors have also suggested that the excessive and prolonged use of oestrogen induces the cancers of endometrium, cervix, vagina, ovary, liver and mammary gland in pre- or post- menopausal women (Cutler *et al.,* 1972; Edmondson *et al.*, 1976; Gambrell *et al.*, 1983; Loose and Stancel, 2006; Madhuri, 2008; Madhuri *et al.*, 2007b; Meissner *et al.*, 1957; Satoskar *et al.*, 2005; Shar and Kew, 1982; Vessey *et al.*, 1983).

In the ovarian cancer study, American Cancer Society researchers reported that 0.2 per cent pre- and post- menopausal women had died from ovarian cancer. Among the women who died, about 32 per cent had taken ORT. The study concluded that women taking oestrogen for 6 years or more had a 40 per cent greater risk of dying from ovarian cancer, and the women taking oestrogen for at least 11 years had a 70 per cent greater risk. However, the ovarian cancer is a rare disease, but its survival rates are low (Madhuri *et al.*, 2007b).

One of the most potent oestrogens, EO in doses of 250, 500 and 750 µg per kg, orally, weekly for 8 and 12 weeks, has been reported to cause cytotoxicity in uterus and ovary (Madhuri, 2008; Madhuri *et al.*, 2007b), and liver (Pandey *et al.*, 2009 and 2011) of rats. Thus, a detailed study was done (Madhuri *et al.*, 2007b), as given herein, to assess the cancerous effects of EO (oestrogen) on the uterus and ovary of albino rats, so as to produce the oestrogen induced cancer models of uterus and ovary, for the first time in India, for the oncological, pharmacological, toxicological and other biological researches.

Experimental Study

Twenty-four healthy inbred female albino rats (each weighing 100-150 g) were undertaken in the study. They were kept in polypropylene cages under standard laboratory conditions of photoperiod (10 hr light:14 hr darkness), temperature ($25^0 \pm 5^0$C) and relative humidity (RH 45-55 per cent) in the animal house of the research place (College of Veterinary Science and Animal Husbandry, Jabalpur, Madhya Pradesh, India). The rats were fed on standard pellet diet and clean drinking water. However, the animals were fasted overnight before the experiment, but water was given *ad libitum*. The experimental designs and protocol in the study received the approval of *'Institutional Animal Ethics Committee'* (IAEC). For production of uterine and ovarian cancer models, the Lynoral tablets (containing 0.05 mg of EO only in each tablet) were triturated and suspended in distilled water mixed with a pinch of *Gum acacia* powder (since the drug is insoluble in water). Sodium chloride (NaCl) solution (0.9 per cent) was used as saline (normal saline). During experiments, the required quantity of saline was also mixed with a pinch of *Gum acacia* powder to have uniformity with the EO suspension. Other chemicals and reagents required for histopathological study were purchased from the chemical shops.

The female albino rats were divided were divided into four groups, each had six rats. The EO was administered to the rats of groups 2, 3 and 4, respectively at the dose of 250, 500 and 750 µg per kg, orally, weekly for 20 weeks. However, the rats of group 1 were given saline to serve as control. To assess the cancerous effects of EO, the rats of group 1 were sacrificed on the 1st week, whereas the rats of groups 2, 3 and 4 were sacrificed on the 21st week of the experiment. The uteri and ovaries of rats were collected and preserved in 10 per cent buffered formalin. Then the uterine and ovarian tissues were processed and stained with Harris's haematoxylin and eosin (H&E) stain as per the method cited by Culling (1963). Thereafter, the uterine and ovarian tissues of the rats of all groups were examined microscopically to assess the uniform and optimum cancerous changes.

On histopathological examination at the 21st week, the uterine tissues of the rats of group 2 (EO @ 250 µg/kg), group 3 (EO @ 500 µg/kg; Figure 4) and group 4 (EO @ 750 µg/kg; Figure 5) appeared with varying degrees of congestion, degeneration and necrosis in the mucosa and endometrial gland with infiltration of inflammatory cells. In group 4, the endometrial glands were severely necrosed and the eosinophil infiltration was quite conspicuous. Proliferation of endometrium in the form of papillary projections was noticed (Figure 5), which indicated the hyperplasia of endometrium. At places, interestingly, hyperplastic epithelium appeared to invade the basement membrane also, suggestive of cancerous (malignant) growth. Histopathologically, the ovarian tissues of the rats of groups 2, 3 (Figure 6) and 4 (Figure 7) at the 21st week revealed marked congestion with vascular wall thickening, fibrosis, degeneration and necrosis in follicular tissues. The changes were found to be more severe and extensive in group 4. Interestingly, the papillary proliferation in the surface epithelium and hyperplasia of follicular cells were seen (Figure 7), which indicated the development of cancer (Madhuri *et al.*, 2007b).

The uterine and ovarian histopathology similar to this study has been observed by many authors (Hertz, 1976; Loose and Stancel, 2006; Madhuri, 2008; Platt, 2005; Satoskar *et al.*, 2005). Various cancerous changes in the uterus of rabbits were observed by Meissner *et al.* (1957) after the administration stilbestrol (a synthetic oestrogen). Adenosquamous carcinoma of the endometrium in women chronically treated with stilbestrol for gonadal dysgenesis has been observed by Cutler *et al.* (1972). Furthermore, Platt (2005) cited that the excessive intake of oestrogen is troublesome and may cause many illnesses, including polycystic ovary syndrome and ovarian cytotoxicity. Thus, the long-term use of oestrogen has been reported to promote the development of endometrial cancer. In earlier experiments, the EO (@ 250, 500 and 750 µg/kg, orally, weekly for 8, 12 and 16 weeks) have been reported to cause uterine and ovarian cytotoxicity in rats (Madhuri, 2008).

In conclusion, the EO (@ 750 µg/kg, orally, weekly) caused severe uterine and ovarian damage, leading to development of optimum (standard) cancer. The extent and severity of damage were dose dependent, suggesting that at higher dose for prolonged period, oestrogen can cause gynecologic (uterine and ovarian) cancer. This study is probably new work in India since no Indian literature could be traceable with regard to oestrogen induced experimental gynecologic cancer in rats. This experiment has important role in knowing the standard toxic dose and duration of oestrogen used as OC and HRT by the millions of women. Hence, the women must be warned against the cancerous effect of oestrogen. The cancer models created through this study can also be applied to produce standard carcinogenesis/cancer in the uterus, ovary, vagina, etc. that will be helpful for evaluating anticancer activity of many drugs (Madhuri *et al.*, 2007b).

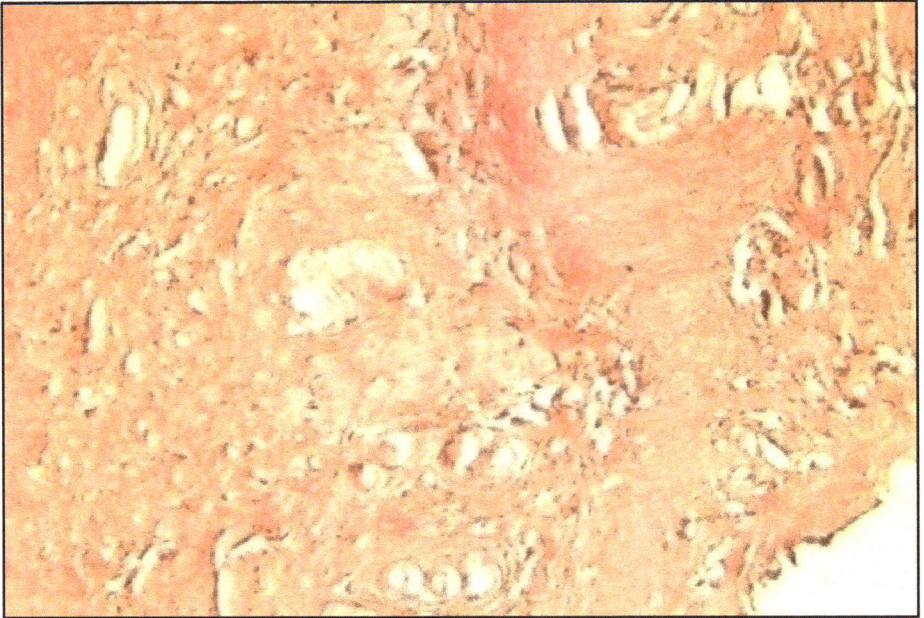

Figure 4: Uterus of Rat Dosed with EO (500 µg/kg, orally, weekly for 20 weeks) Shows Varying Degrees of Congestion, Degeneration and Necrosis in Mucosa and Endometrial Gland with Infiltration of Inflammatory Cells (H&E, x100) [Madhuri, 2008; Madhuri *et al.,* 2007b].

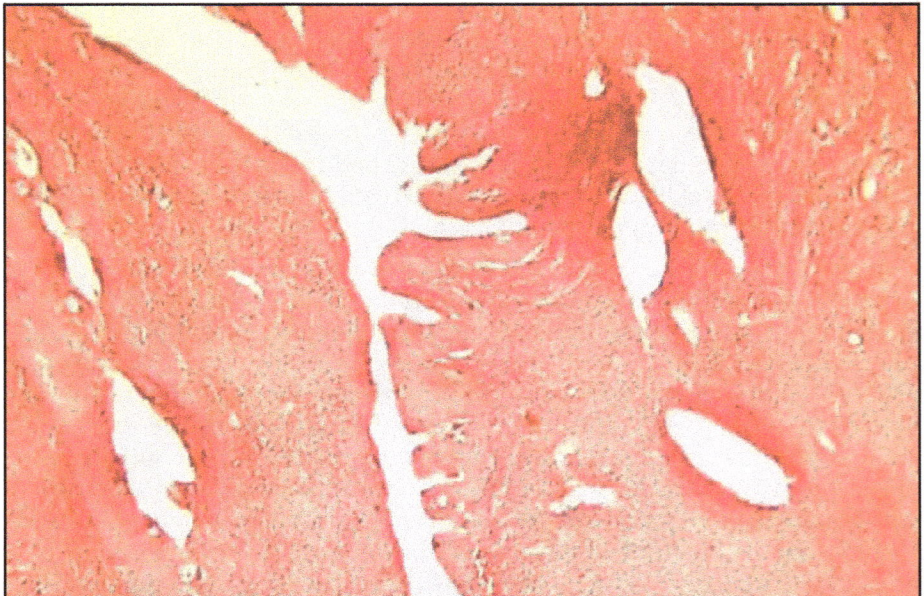

Figure 5: Uterus of Rat Dosed with EO (750 µg/kg, orally, weekly for 20 weeks) Shows Hyperplasia (as papillary projections) of Endometrium. At places, hyperplastic epithelium appears to invade the basement membrane, indicating the cancerous growth (H&E, x100) [Madhuri, 2008; Madhuri *et al.,* 2007b].

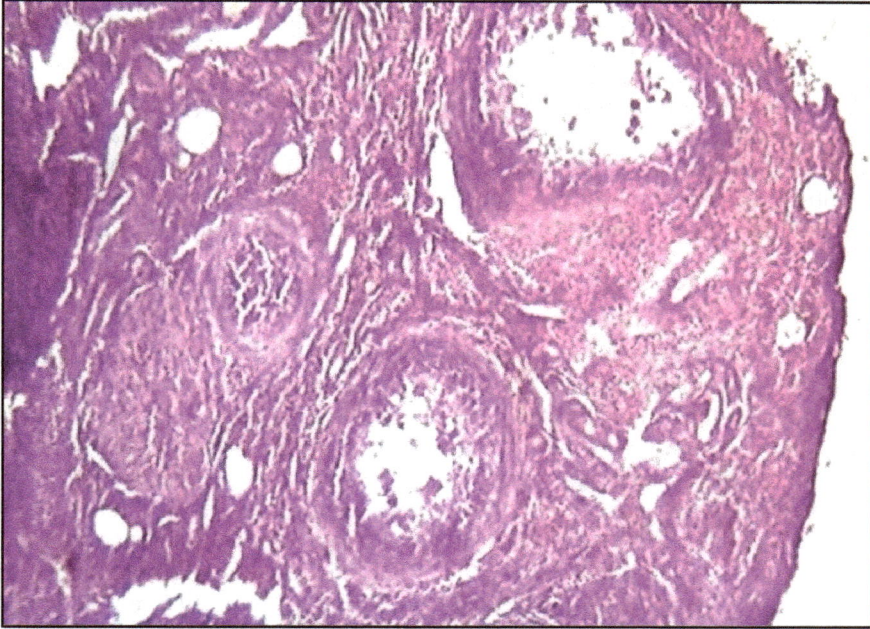

Figure 6: Ovary of Rat Dosed with EO (500 µg/kg, orally, weekly for 20 weeks) Shows Extensive Fibrosis, Degeneration and Necrosis of Follicular Tissue with Extensive Infiltration of Lymphocytes in the Interstitial Tissues (H&E, x400) [Madhuri, 2008; Madhuri *et al.*, 2007b].

Figure 7: Ovary of Rat Dosed with EO (750 µg/kg, orally, weekly for 20 weeks) Shows Papillary Proliferation in Surface Epithelium and Hyperplasia of Follicular Cells, indicative of the development of cancer (H&E, x400) [Madhuri, 2008; Madhuri *et al.*, 2007b].

Cytotoxicity Leading to Cancer Caused by Estrogen in Uterus of Albino Rat

EO estrogen has been reported to cause cytotoxicity leading to cancer in various organs, including uterus, ovary and liver (Madhuri, 2008). The median lethal dose (LD_{50}) of EO has been determined to be more than 1000 µg per kg body weight, orally in female albino rats (Pandey and Madhuri, 2008a). In this view, an experiment was conducted to assess the uterine cytotoxicity leading to cancer due to administration of EO in rat.

Experimental Study

Thirty-six healthy inbred female albino rats (100-150 g) were equally divided into six groups (each group had six animals). The rats were kept in polypropylene cages under standard laboratory conditions in the animal house, and fed on standard pellet diet and drinking water. However, the animals were fasted overnight before the experiment, but water was given *ad libitum*. The experimental designs and protocol in the study received the approval of IAEC. The Lynoral tablets (containing 0.05 mg of EO only in each tablet) were triturated and suspended in distilled water mixed with a pinch of *Gum acacia* powder (since the drug is insoluble in water). During experiments, the required quantity of normal saline was also mixed with a pinch of *Gum acacia* powder to have uniformity with the EO suspension. Other chemicals and reagents required for histopathological study were purchased from the chemical shops.

The EO was administered at the rate of 750 µg per kg body weight, orally, weekly for 8, 12, 16, 20 and 24 weeks in groups 2, 3, 4, 5 and 6 of female albino rats, respectively. However, the rats of group 1 were given normal saline (0.9 per cent solution of NaCl) to serve as control. After the experimental periods, the rats of different groups were sacrificed, and their uteri were collected and preserved in 10 per cent buffered formalin. Later on, the uterine tissues were processed and stained with Harris's H&E stain as per the method described by Culling (1963). To assess the histopathological cancerous changes, the uterine tissues were observed, microscopically.

Microscopic examination of H&E stained section of the uterus of group 1 showed normal histological profile. On the 9th week (Group 2; Figure 8), the vacuolar degeneration, necrosis and desquamation of columnar epithelial cells in the mucosa and endometrial glands were observed. The blood vessels in the endometrium were dilated and congested. At certain places, proliferation of fibrovascular connective tissues and infiltration of inflammatory cells were distinct. On the 13th week (Group 3; Figure 9), marked vascular congestion, epithelial necrosis and fibrous tissue proliferation were seen. The fibrosis was extensive, resulting into compression of endometrial glands. Desquamation of glandular epithelium was also noticed. On the 17th week (Group 4; Figure 10), proliferation of endometrium in the form of papillary projections lined with tall columnar cells was seen. Hyperaemia of blood vessels was also observed. The endometrial glands were necrosed and conspicuous eosinophil infiltration was noticed. On the 21st week (Group 5; Figure 11), focal hyperplasia of endometrial lining in the form of papillary projections (as endometrioid mass) and

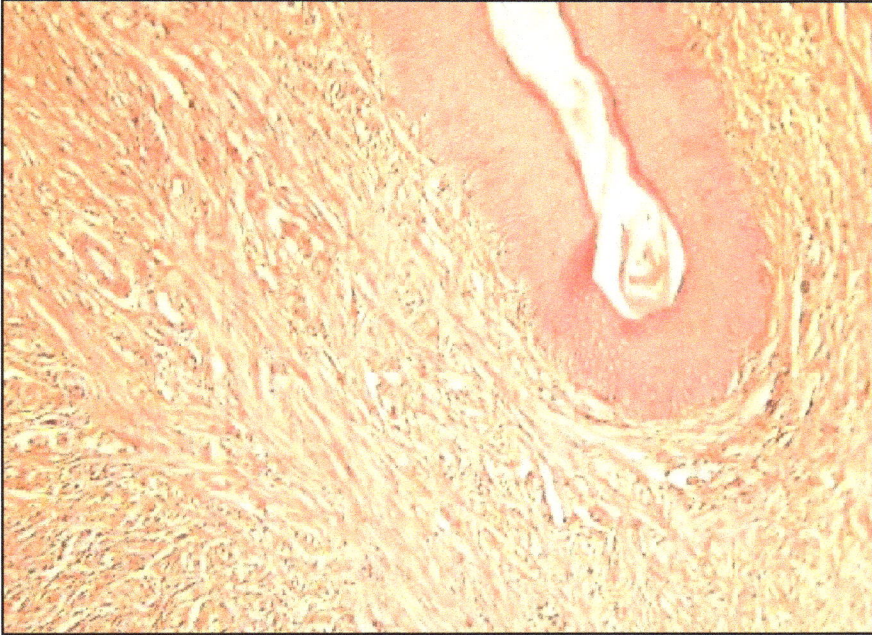

Figure 8: Uterus of Rat Administered with EO (750 µg/kg, orally, weekly for 8 weeks) Reveals Necrosis, Desquamation of Epithelial Dells, Infiltration of Inflammatory Cells and Proliferation of Fibrovascular Connective Tissues (H&E, x400) [Madhuri, 2008].

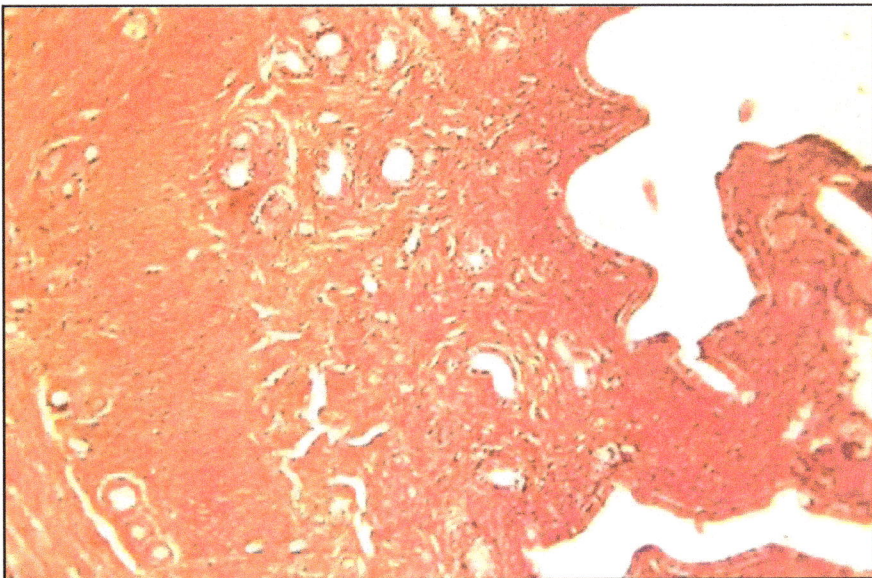

Figure 9: Uterus of Rat Administered with EO (750 µg/kg, orally, weekly for 12 weeks) Reveals Marked Vascular Congestion, Epithelial Necrosis and Severe Fibrosis, Leading to Compression of Endometrial Glands (H&E, x100) [Madhuri, 2008].

Treatment of Cancers by Medicinal Plants

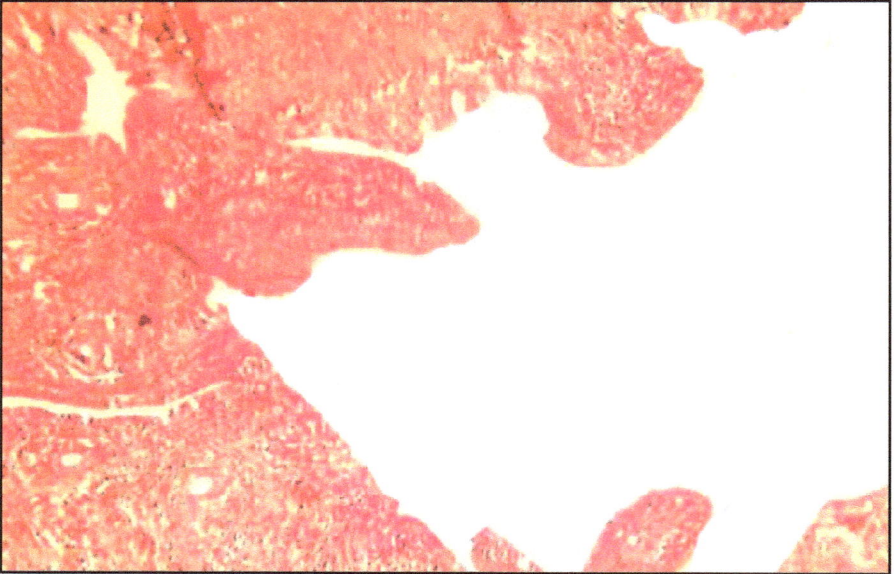

Figure 10: Uterus of Rat Administered with EO (750 μg/kg, orally, weekly for 16 weeks) Reveals Proliferation of Endometrium in the Form of Papillary Projections Lined with Tall Columnar Cells (H&E, x100) [Madhuri, 2008].

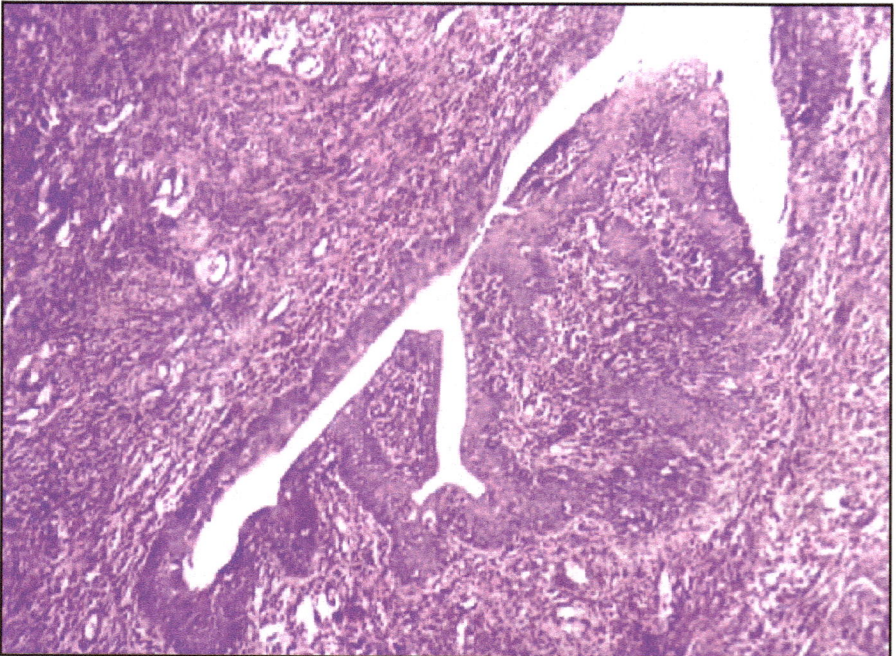

Figure 11: Uterus of Rat Administered with EO (750 μg/kg, orally, weekly for 20 weeks) Reveals Hyperplastic Epithelium, Indicating the Carcinogenesis (x100, H&E) [Madhuri, 2008].

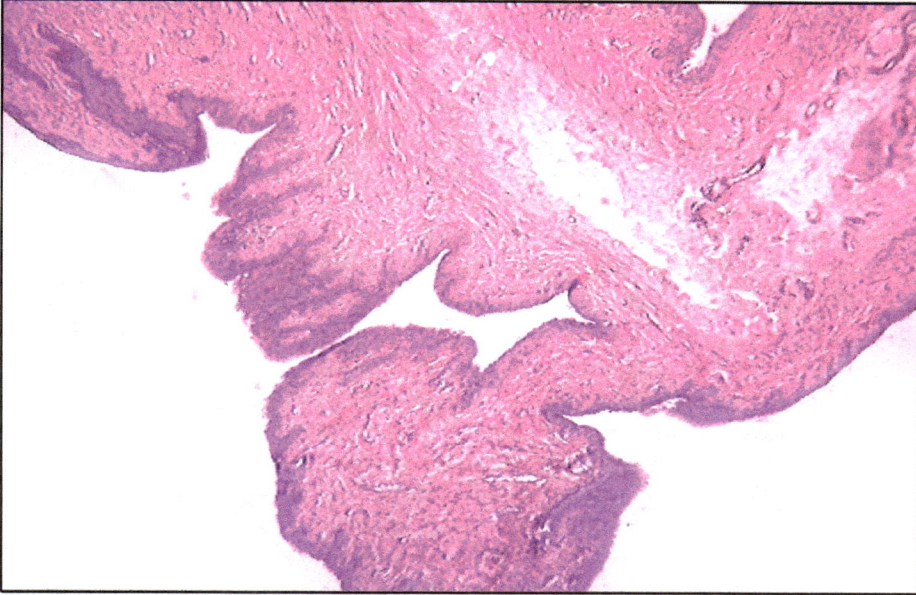

Figure 12: Uterus of Rat Administered with EO (750 µg/kg, orally, weekly for 24 weeks) Reveals Extensive Focal Hyperplasia of Endometrial Lining (epithelium) (H&E, x100) [Madhuri, 2008].

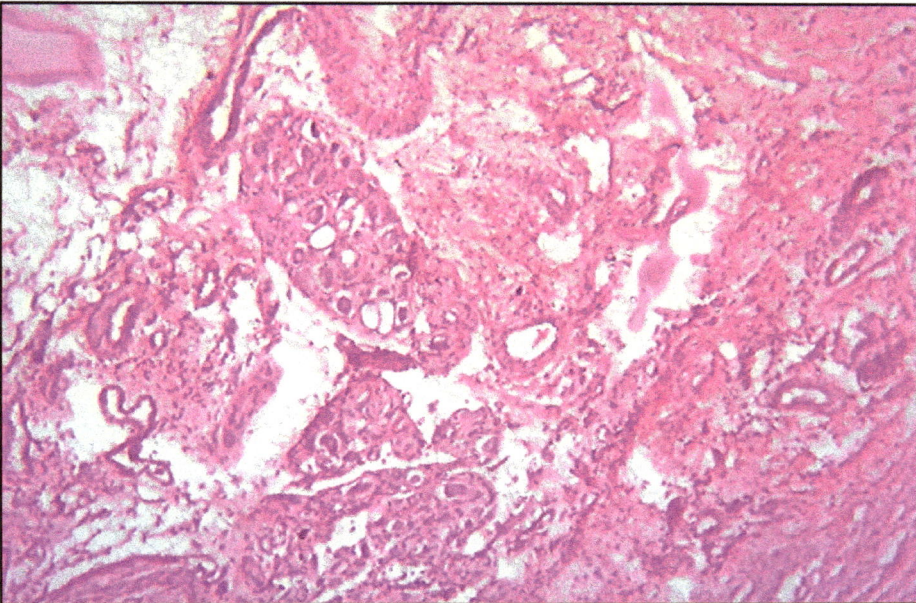

Figure 13: Uterus of Rat Administered with EO (750 µg/kg, orally, weekly for 24 weeks) Reveals many Malignant Changes including Hyperchormasia, Enlarged Nuclei, Anisokaryosis, Anisocytosis, New Angiogenesis and Hlandular Polarity, Indicating the Uterine Carcinogenesis or Adenocarcinoma; Disarray, Leading to Severe Malignancy in the Whole is seen (H&E, x100) [Madhuri, 2008].

fibrovascular connective tissues were seen. At places, interestingly, the hyperplastic epithelium appeared to invade the basement membrane also, indicating the cancerous growth. On the 25th week (Group 6; Figures 12 and 13), extensive focal hyperplasia of endometrial lining (epithelium) was seen. Various other malignant changes such as hyperchormasia, enlargement of nuclei, anisokaryosis, anisocytosis and glandular polarity were seen; disarray, leading to severe malignancy in the whole area. All these changes in group 5 suggested the development of endometrial cancer or endometrial adenocarcinoma. The extent and severity of uterine damage were time dependent, as EO after prolonged period (20-24 weeks) caused severe changes.

The results of this study may be correlated with the reports of various authors (Hertz, 1976; Loose and Stancel, 2006; Meissner *et al.*, 1957), who cited that the cytotoxicity, leading to cancer in the uterus of human beings and animals may occur after administration of estrogen. Jabara (1962) reported that the endometrium of bitches after chronic dosing of stilbestrol showed degeneration with marked glandular atrophy. Schwartz *et al.* (1969) observed that high and prolonged dose of quinestrol estrogen caused uterine enlargement with endometrial hyperplasia and myometrial hypertrophy in bitches. The uterine adenocarcinoma was noticed by Newbold and Liehr (2000) in mice after 12- and 18- months dosing of EO and some other estrogens (*viz.*, 2- and 4-hydroxyestradiol, DES and estradiol). However, probably no research work on estrogen induced experimental uterine cytotoxicity has been conducted in India as no Indian literature could be available in this regard.

A Model of Estrogen Induced Ovarian Adenocarcinoma

The OCs are preferred for the contraception, and over 100 million women worldwide use OC steroids (*e.g.*, estrogen preparations) every year; however, many of them are carcinogenic. The ORT is associated with risk of epithelial ovarian cancer of the endometrioid type (Madhuri and Pandey, 2010b). An estrogen, EO in doses of 250, 500 and 750 µg per kg, orally, weekly for 8 and 12 weeks caused cytotoxicity in the uterus (Madhuri *et al.*, 2009) and ovary (Madhuri *et al.*, 2007c) of rats.

Hence, a study was conducted (Madhuri and Pandey, 2010b), as detailed here, to produce a model of ovarian adenocarcinoma by administering the EO (estrogen) in rat. This is probably the first research work in India to produce ovarian cancer by EO in the rat since no Indian literature other than this could explore such research work.

Experimental Study

Twenty-four healthy inbred female albino rats (each weighing 100-150 g) were divided into four groups (each group consisted of 6 rats). The animals were kept under standard laboratory conditions in animal house and fed on standard pellet diet and clean drinking water. However, the animals were fasted overnight before the experiment, but water was given *ad libitum*. The experimental designs and protocol in the study received the approval of IAEC. The Lynoral tablets (containing 0.05 mg of EO only in each tablet) were triturated and suspended in distilled water mixed with a pinch of *Gum acacia* powder (since the drug is insoluble in water). During experiments, the required quantity of normal saline was also mixed with a pinch of *Gum acacia* powder to have uniformity with the EO suspension. Other chemicals and reagents required for histopathological study were purchased from the chemical shops. The EO was administered at the rate of 750 µg per kg body weight, orally, weekly for 16, 20 and 24 weeks in groups 2, 3 and 4 of female albino rats, respectively. However, the rats of group 1 were administered saline to serve as control. After the end of experiments, the rats were sacrificed and the ovaries were preserved in 10 per cent buffered formalin. The ovarian tissues were processed and stained with Harris's H&E stain as per the method described by Culling (1963). On microscopic examination, the cancerous changes in the rat ovaries were assessed by observing the uniform and optimum damage.

On the 17th week of the administration of EO in rats (Group 2), the ovarian tissues revealed degeneration, necrosis and fibrosis of follicular tissues, extensive infiltration of lymphocytes in interstitial tissues, thickening of the vascular wall, and presence of homogeneous mass in the lumen (Figure 14). On the 21st week (Group 3), the ovarian tissues revealed more severe and extensive pathological lesions than those noticed in group 2. Fibroplasia of interfollicular connective tissues, hyperplasia of follicular cells and papillary proliferation in surface epithelium were observed (Figure 15). On the 25th week (Group 4), besides these changes, other malignant lesions, such as hyperchromatosis, enlargement of nuclei, anisokaryosis and

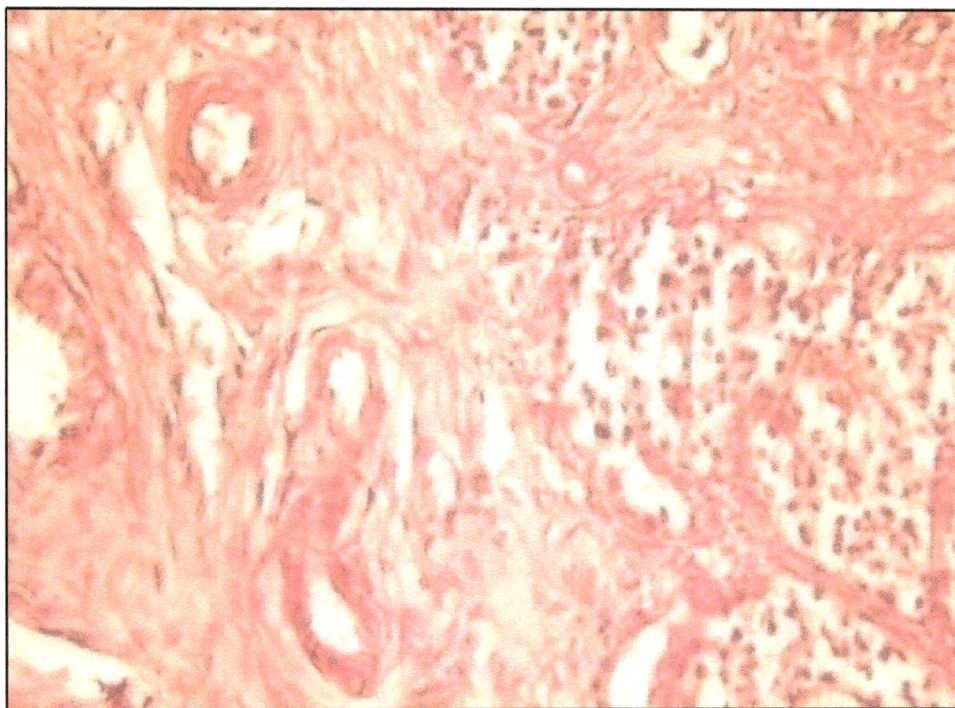

Figure 14: Ovary of Rat on 17th Week of EO (750 µg/kg, orally, weekly for 16 weeks) Administration, Shows Necrosis and Fibrosis of Follicular Tissues, and Extensive Infiltration of Lymphocytes in Interstitial Tissues (H&E, x400) [Madhuri, 2008; Madhuri and Pandey, 2010b].

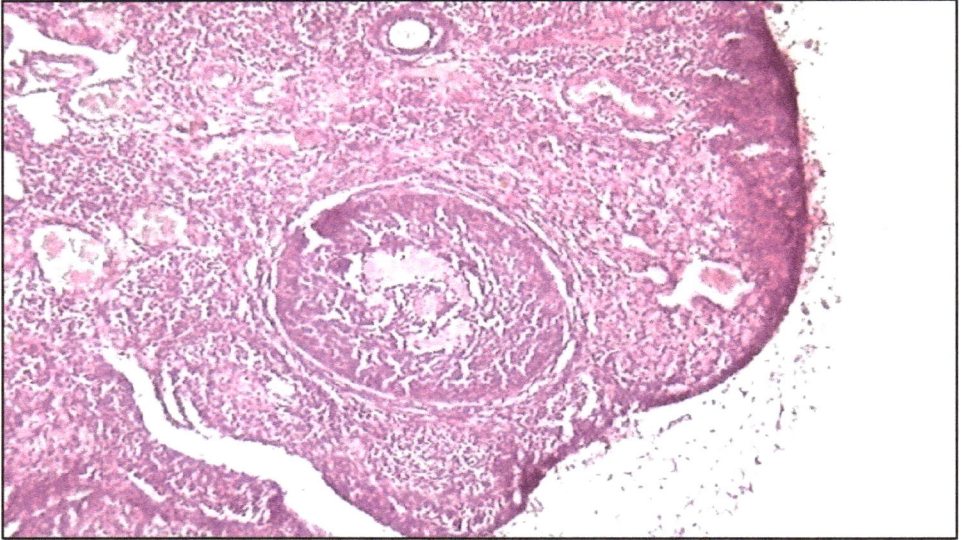

Figure 15: Ovary of Rat on 21ˢᵗ Week of EO (750 μg/kg, orally, weekly for 20 weeks) Administration, Shows Fibroplasia of Interfollicular Connective Tissues (H&E, x100) [Madhuri, 2008; Madhuri and Pandey, 2010b].

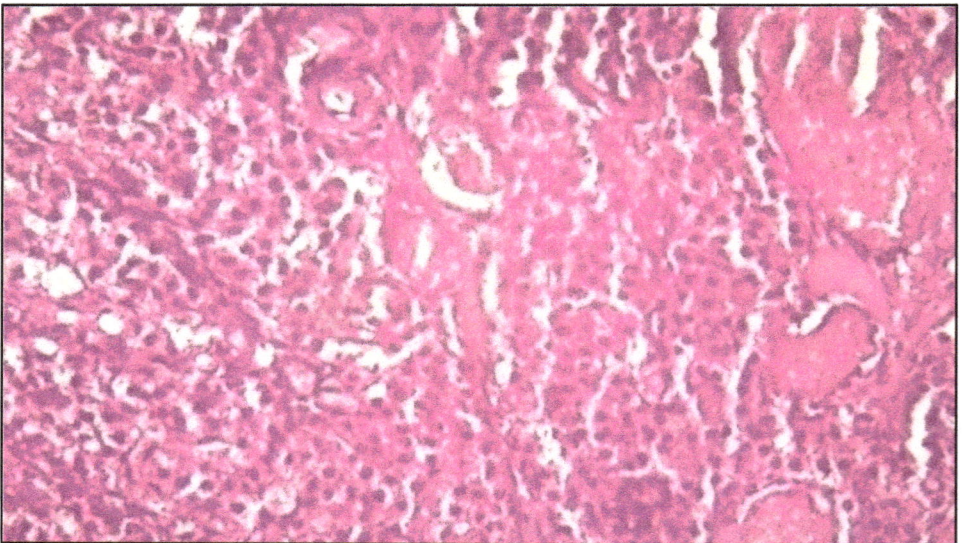

Figure 16: Ovary of Rat on 25ᵗʰ Week of EO (750 μg/kg, orally, weekly for 24 weeks) Administration, Shows Malignant Lesions such as Hyperchromatosis, Enlargement of Nuclei, Anisocytosis and Anisokaryosis (H&E, x100) [Madhuri, 2008; Madhuri and Pandey, 2010b].

anisocytosis were seen (Figure 16). All these severe malignant changes indicated the presence of ovarian adenocarcinoma. The extent and severity of ovarian damage were time dependent, which suggested that at higher dose for prolonged period, EO may develop the cancer in ovary.

The above results may be correlated with the reports of earlier authors (Hertz, 1976; Loose and Stancel, 2006; Platt, 2005). Various carcinogenic changes in rabbit ovary were seen by Meissner *et al.* (1957) after the administration of estrogen. Ishimura *et al.* (1986) reported that estradiol may cause tumours arising from interstitial cell hyperplasia, with proliferation and invasion of the ovarian surface epithelium into the ovarian stroma. After continuous exposure of estradiol, Bai *et al.* (2000) and Silva *et al.* (1998) have noticed the formation of a papillary ovarian surface, resembling human serous neoplasms of low malignant potential in laboratory animals; while Murdoch and Van Kirk (2002) have observed the ovarian surface epithelial cell proliferation in sheep.

Estrogen Caused Hyperplasia and Cancer in Rat Ovary

Of course, *'estrogen is the necessity of female because without estrogen female is no more female; however, excessive and prolonged use of estrogen may cause cancers of many organs'* (Madhuri and Pandey, 2011). EO has damaged the hepatic tissues of rats, when administered at the dose of 500 µg per kg, orally, weekly for 8 and 12 weeks (Pandey *et al.*, 2008). With this view, an experiment was done (Madhuri and Pandey, 2011) to assess the cancerous effect of EO at a specific dose and period on the rat ovary.

Experimental Study

Forty-two healthy inbred female albino rats (100-150 g) were divided into seven groups, each had six rats. The animals were kept in polypropylene cages under standard laboratory conditions with $25^0\pm5^0$C temperature, 45–55 per cent RH and 10 hr light:14 hr darkness in the animal house, and fed on standard pellet diet and clean drinking water. However, before experiment the animals were fasted overnight, but water was given *ad libitum*. The experimental designs in the study received approval of IAEC. The Lynoral tablets (containing 0.05 mg of EO only in each tablet) were triturated and suspended in distilled water mixed with a pinch of *Gum acacia* powder (since the drug is insoluble in water). During experiments, the required quantity of normal saline was also mixed with a pinch of *Gum acacia* powder to have uniformity with the EO suspension. Other chemicals and reagents required for histopathological study were purchased from the chemical shops. The rats of groups 2, 3 and 4 were administered with EO at the rate of 250, 500 and 750 µg per kg, orally, weekly for 16 weeks, respectively. The same doses of EO were administered to the rats of groups 5, 6 and 7, respectively for 20 weeks. However, the rats of group 1 were administered normal saline to serve as control. After the respective experimental period, the rats were sacrificed and their ovaries were collected and preserved in 10 per cent buffered formalin. Thereafter, the ovarian tissues were processed and stained with Harris's H&E stain as per the method cited by Culling. On microscopic examination, the pre-cancerous and cancerous lesions in the ovaries of rats of groups 2 to 7 as compared to group 1 (control) were observed.

On the 17th week, the ovaries of group 2 (EO @ 250 µg/kg) and group 3 (EO @ 500 µg/kg) showed marked congestion and fibrosis. Degeneration and necrosis of follicular tissue were also noticed. In group 4 (EO 750 µg/kg, Figure 14), the histopathological changes were more marked. Presence of homogenous mass in the lumen, infiltration of lymphocytes and thickening of blood vascular walls were quite conspicuous. On the 21st week, the histopathological changes were more severe in groups 5 to 7, subsequently. In group 6 (EO @ 500 µg/kg, Figure 6), extensive fibrosis, including severe degeneration and necrosis of follicular tissues were observed. In group 7 (EO @ 750 µg/kg, Figures 7 and 15), these changes were quite conspicuous with fibroplasia of interfollicular connective tissues. Interestingly, hyperplasia of follicular cells, papillary proliferation in surface epithelium and other malignant changes were observed, indicating the ovarian carcinogenesis. The extent and severity of ovarian damage were dose and time dependent, suggesting that the EO at higher dose for prolonged period (750 µg/kg, orally, weekly for 20 weeks) may cause hyperplasia leading to cancer in the ovary. The results of this study are coined with various reports of ovarian cancer caused by oestrogen (Hertz, 1976; Loose and Stancel, 2006; Madhuri, 2008; Rossing *et al.*, 2007).

Chapter 3

Anticancer Medicinal Plants in General

History and Importance

Plants have been in use for treating different body disorders from the prehistoric times and continue to be the source of more than 25 per cent of the present range of prescribed drugs. In the indigenous system of medicine all over the world, for over 2000 years, plants have been used against many kinds of cancer. The earliest record of herbal treatment can be traced to ancient Chinese and Greek texts (Evans, 1989). The Unani and Ayurvedic systems have also used a huge number of plants for the treatment of cancer (Pandey *et al.,* 2013a).

In spite of the spectacular advances made by Medical Science during the present century, treatment of cancer remains an enigma. From the time of Galen (ca. A.D. 180), the juice of woody nightshade (*Solanum dulcamara*) has been used to treat various tumours (including cancers), and warts. The main active tumour inhibitor has been identified as the steroidal alkaloid, glycoside β-solamarine. In folk medicine, even various lichens, *e.g., Cladonia, Cetraria* and *Usnea* have a history of their application cancers. These plants are rich sources of usnic acid, a compound recognized for many years as antibacterial and antifungal, but only recently, as an antitumour compound. Similarly, many centuries ago, the Druids claimed that mistletoe (*Viscum album*) could be used to cure cancer. From the extract of *V. album*, 11 protein fractions with significant antitumour activity have been isolated. Mezeron (*Daphne mezereum*), despite its toxic properties, has also been recommended in many countries. The active antitumour constituents of this plant have been identified as a diterpene derivative, mezerin. However, prior to 1940s, the chief non-surgical treatment of cancer was X-ray and radium therapy, besides arsenicals and urethane, applied locally to destroy the inflicted tissues. During the 1940s, the radioisotopes, nitrogen mustards, sex hormones and antifolic acidic agents were developed. In recent years, efforts have been made to synthesize potential anticancer drugs. Consequently, hundreds of chemical variants of known classes of cancer therapeutic agents have been synthesized. It is recognized that a successful anticancerous drug should be the one which kills or incapacitates cancer cells without causing excessive damage to the normal cells. This criteria is difficult, or perhaps, impossible to attain, and that is why cancer patients suffers unpleasant side effects while undergoing treatments (Shrivastava *et al.*, 2002).

Plants have been a source of medicinal agents since time immemorial. From the dawn of civilization, men have been utilizing the important biological properties of various plants for the treatment of different diseases. Chemists and biologists have been actively engaged for a long time to discover potent drugs from natural source for combating cancer. More than 1,000 plants possess significant anticancer properties. Hence, plants are the most vital source of many compounds, which have significant therapeutic values for combating cancer (Rajan and Chezhiyan, 2002).

Worldwide research is going on to find out the effective anticancer drugs. Allopathic system of medicine (chemotherapy, radiation, surgery, etc.) has several untoward effects. *'Traditional Indian Medicine/Indigenous System of Medicine'* has many indigenous (herbal) formulations with versatile medicinal properties. Plant-based drugs have become popular throughout the world nowadays, also for cancer therapy without causing any toxicity. India is the largest producer of medicinal plants and is rightly called the *'Botanical Garden of the World'*. In India, out of 45,000 plant species identified, about 15 to 20 thousand plants are of good medicinal value. The traditional communities use only about 7000 to 7500 plants for medicinal purpose; the Siddha system of medicine uses about 600, Ayurveda 700, Unani 700 and modern medicine about 30 medicinal plants for treating variety of diseases. However, only few medicinal plants have attracted the scientists to investigate the remedy of cancer (Madhuri, 2008; Pandey, 2007; Pandey *et al.*, 2013a; Somkuwar, 2003). On the other hand, there are about 400 families of flowering plants, out of which at least 315 are represented by India (Pandey and Madhuri, 2008c). As per another report (Pandey *et al.*, 2013a), the medicinal plants comprise about 8000 species and account for approximately 50 per cent of all the higher flowering plant species of India. Medicinal values of few such plants have been reported but still a good number of plants are to be explored out. Kathiresan *et al.* (2006) stated that although more than 1500 anticancer drugs are in active development with over 500 of drugs under clinical trials, there is a definite need to develop more effective and less toxic anticancer drugs.

According to the WHO, about three-quarters of the world's population currently use herbs and other forms of traditional medicines to treat several diseases. The traditional medicines are widely used in India as well. Even in the United States of America (USA), the use of phytomedicines has increased dramatically in the last two decades (Rao *et al.*, 2004). As per the WHO estimates, more than 80 per cent people in developing countries depend on traditional medicine for their primary health needs (Pandey, 2009; Pandey and Madhuri, 2011a; Sivalokanathan *et al.*, 2005). Approximately 14 to 64 per cent of cancer patients use complementary and alternative medicine either as an alternative treatment or an adjunctive treatment to augment conventional therapy (Pandey *et al.*, 2013a).

Henceforth, the *'plant kingdom'* plays a major role in the life of human beings and animals. The plant, as one of the important sources, still maintains its original place in the treatment of various diseases, including tumours (neoplasms), with no ill effects. In fact, ethnomedicinal plants (indigenous system of medicine) are easily available, cheaper and possess no toxicity. The plant kingdom is the milestone to ameliorate the genetic, environmental and malnutrition (food related) carcinogens. Many plants and plant products such as fruits, vegetables, herbs, plant extracts and plant-derived drugs, etc. have been reported for the prevention and treatment of various cancers. Consumption of large amounts of vegetables and fruits can prevent the development of cancer (Madhuri and Pandey, 2009b; Pandey, 2009). Considerable works have been done on such anticancer agents and some of them have been marketed as anticancer drugs, based on the traditional uses and scientific reports (Pandey and Madhuri, 2008b and 2009a; Pandey, 2009). Although mechanism of many anticancer agents is unknown, nevertheless the consumption of such plants and plant products decreases the risk of developing cancer (Pandey and Madhuri, 2008b; Pandey, 2009). A major group of these natural products are the powerful antioxidants, others are phenolic in nature and the remainder includes reactive groups (Pandey, 2009). The medicinal plants, besides having natural therapeutic values against various diseases, provide high quality of food and raw materials for livelihood. In fact, the medicinal plants are easily available, cheaper and possess no toxicity as compared to the modern (allopathic) drugs (Pandey and Madhuri, 2009a).

A great deal of pharmaceutical research has improved the quality of herbal drugs used against various types of cancer. With the advanced knowledge of molecular science and the refinement in isolation and structure elucidation techniques, we are in a much better position now to identify various anticancer herbs. Scientists all over the world are concentrating on the use of herbs to boost immune system of the body against cancer. Scientists have contributed for a number of years to identify hundreds of anticancer herbs, and developed various herbal formulations from their active principles that inhibit growth and spread of cancer without any side effect. Such herbs possess anticancer, immunoenhancing, antiangiogenesis, antioxidant and antimutagenic properties. They inhibit growth and spread of cancer by modulating the activity of hormones, enzymes and other biological factors. The therapeutic effect of these herbs is executed by the complex synergistic interaction among their various active principles (Pandey, 2011b).

The medicinal plants and their products, particularly vegetables have antioxidant activity leading to anticancer effect. Plants used as vegetables prevent human from several diseases, including cancer. The vegetables contain many phytochemicals having antioxidant activity (Pandey and Madhuri, 2011a). The vegetables and fruits have an important role in the treatment of cancer. Many doctors recommend that people wishing to reduce their risk of cancer must eat several pieces of fruits and several portions of vegetables every day (Madhuri and Pandey, 2008; Pandey and Madhuri, 2011a). Intake of 400 to 600 g per day of vegetables and fruits can reduce the occurrence of many common forms of cancers. Thus, the people who eat much quantity of vegetables have about one-half the risk of cancer and less mortality (Madhuri and Pandey, 2008).

Important Anticancer Herbal Drugs and their Mechanism of Action

It is increasingly being realized now that majorities of the diseases are mainly due to the imbalance between pro-oxidant and antioxidant homeostatic phenomenon in the body. Pro-oxidant condition dominates either due to increased generation of free radicals and/or their poor quenching/scavenging into the body. *'Free radicals'* are the fundamental to any biochemical process, and represent an essential part of the aerobic life and our metabolism. They are continuously produced by body, such as respiration and some cell mediated immune functions. There is a dynamic balance between the amount of free radicals generated in the body and antioxidant to quench and or/scavenge them and protect the body against the deleterious effects. Thus, the oxidant status in human reflects the dynamic balance between the antioxidant defense and pro-oxidant conditions and has been suggested as a useful tool in estimating the risk of oxidative damage (Khan, 2009; Madhuri *et al.,* 2011b; Pandey *et al.,* 2013a). Therefore, the activity of *'free radicals'* in our body are naturally controlled and delimited by another group of chemical compounds called *'antioxidants'*, which are present in higher levels in our regularly consumed food materials. The β-carotene, vitamin C and vitamin E (tocopherol) are some important antioxidant agents. They are effective against cancer by scavenging the accumulated free radicals in our body. In normal cells, the ratio of free radicals and antioxidants are strictly balanced (Lakshmi Prabha *et al.,* 2002).

In recent years, immunomodulation has attracted the interest of scientists all over the world to modulate immune system for achieving the objective of preventing an infection rather than treating it at an advanced stage. This led to the introduction of the concept of *'prohost therapy'*, which aims to bolster host immune function to prevent infections. The herbal drugs are known to have good immunomodulatory properties. These drugs act by stimulating both non-specific and specific immunity. The medicinal plants may promote host resistance against infections by re-establishing body equilibrium and conditioning the body tissues. It is now being recognized that immunomodulatory therapy could provide an alternative to conventional chemotherapy for various diseases (Agrawala *et al.,* 2001; Madhuri, 2008; Pandey and Madhuri, 2006b and 2009a; Pandey *et al.,* 2013a). A great variety of allopathic compounds has been investigated, and a few have proven their usefulness in the treatment of neoplasms (tumours). Some immune responses, especially cellular immunity mediated by *'T lymphocytes'*, are also thought to play important roles in the natural resistance of the host against malignant tumours (cancer); others including humoral *'blocking factors'* produced by *'B lymphocytes'*, may be deleterious to the host by interfering with the capacity of cytotoxic lymphocytes to react against neoplastic cells. Coupled with the possibility of genetic damage, the chronic use of immunosupppressive agents carries an increased risk of neoplasia, usually of histiocytic or lymphoid origin. Thus the allopathic antineoplastic drugs which exhibit the most consistent therapeutic activity against certain types of neoplasms are active in dose range which is injurious to normal host tissue (Pandey and Madhuri, 2006a).

Several medicinal plants elicit the anticancer effect due to their potent *'antioxidant'* activities. The main *'antioxidant-phytochemicals'* of the plant-drugs include vitamins (A, C, E and K), carotenoids (carotene), terpenoids, flavonoids (quercetin, flavones, isoflavones, flavonones, anthocyanins, catechins and isocatechins), polyphenols (ellagic acid, gallic acid and tannins), alkaloids, glycosides, polysaccharides, saponins, lignins, xanthenes, enzymes (SOD, catalase and glutathion peroxidase) and minerals [selenium (Se), Cu, Mn, Zn, Cr and iodine (I)], etc. All these antioxidants prevent the humans as well as animals from the cancer and other diseases by protecting the cells from *'oxidative damage'* caused by *'free radicals'* - the highly reactive oxygen compounds (Gupta and Sharma, 2006; Kathiresan *et al.*, 2006; Madhuri, 2008; Madhuri and Pandey, 2008 and 2009a; Pandey and Madhuri, 2009a and 2011a; Pandey *et al.*, 2013a; Polidori, 2003; Ray and Hussan, 2002; Vecchia and Tawani, 1998). The mechanism of action of plant-drugs can be understood by few examples, as discussed here.

The most successful higher plant materials used in cancer chemotherapy are the alkaloids of *Catharanthus roseus* (Apocynaceae family) in which a number of dimeric alkaloids having antileukaemic properties are reported. Vinblastine and vincristine are the two indole-dihydronidole alkaloids. These are typical representative of *C. oncolytic* alkaloids, and have been developed as commercial drugs (Shrivastava *et al.*, 2002). In fact, *C. roseus* G. Don (formerly called *Vinca rosea* Linn., also known as *Lochnera rosea* Linn. Reichb. and 'Sadabahar' in Hindi) contains four active alkaloids: vinblastine, vincristine, vinleurosine and vinrosidine. Two of these, vinblastine and vincristine, are important clinical agents for the treatment of leukemias, lymphomas and testicular cancer. Another alkaloid, vinorelbine, has important activity against lung and breast cancers. Vindesine, semisynthetic derivative of vinca alkaloid, is useful in treating the neoplasms including lymphomas, blastic crisis of chronic granulocytic leukemia and systemic mastocytosis. The vinca alkaloids are cell-cycle-specific agents which block mitosis with metaphase arrest; the cells blocked in mitosis undergo changes characteristic of apoptosis. They bind specifically with the protein tubulin, a key component of cellular microtubules. However, vinca alkaloids cause cytotoxicity in normal cells (Chabner *et al.*, 2006; Pandey and Madhuri, 2006a; Pandey *et al.*, 2013a).

Podophyllum hexandrum and *P. peltatum* (Podophyllaceae plant family) are used by the ancient Chinese as antitumour drugs. Thirty lignans from these plants exhibit the cytostatic or antitumour properties (Chabner *et al.*, 2006; Shrivastava *et al.*, 2002). *P. hexandrum* Royle (*P. emodi* Wall. ex Hook. f. and Thoms.) contains podophyllin resin and podophyllotoxin as active principles. Podophyllin acquires importance because of its possible use in controlling some forms of cancer (CSIR, 1986). The major active compounds isolated from *Podophyllum* are podophyllotoxin (extracted from roots and rhizomes of *P. peltatum*, mandrake plant or May apple), α-peltatin and β-peltatin. Both podophyllin and podophyllotoxin experimentally showed remarkable antitumour activity, but these could not stand clinical trials due to their toxic effects. Hence, the derivatives of podophyllotoxin (active principle) with less toxicity and more therapeutic efficacy were discovered as the two semisynthetic glycosides (epipodophyllotoxins) *viz.*, VM-26 (teniposide) and VP16-213 (etoposide). Both these showed antitumour activity against many tumours including small cell lung carcinomas, testicular tumours, Hodgkin's disease, paediatric leukemia and large cell lymphomas. Epipodophyllotoxins (etoposide and teniposide) at low concentrations block cells at the S-G_2 interphase of cell cycle and, at higher concentrations, cause G_2 arrest. Single-strand DNA breaks are seen in intact cells but not with purified DNA, suggesting that cellular enzymes are in some way involved. Thus, the podophyllotoxin and its analogues appear to exert their cytostatic activity by at least two different mechanisms. In the first one, *e.g.*, podophyllotoxin inhibits the microtubule assembly. In the second mechanism, the etoposide interacts with DNA topoisomerase II. The podophyllotoxin and α-peltatin show 0 per cent and 20 per cent inhibition of DNA topoisomerase II, respectively but etoposide and teneposide show more than 88 per cent of this inhibition, though they do not inhibit microtubule polymerization. The latter two affect the cell division in the late S and G_2 phase of cell cycle. This is connected with enzyme-mediated DNA cleavage which leads to tumour cell death (Chabner *et al.*, 2006; Pandey and Madhuri, 2006a; Pandey *et al.*, 2013a; Shrivastava *et al.*, 2002).

Camptothecin, a novel pyrrolo (3,4-B) quinolone alkaloid with significant anticancer activity has been isolated from *Camptotheca acuminata* Decne (Nyssaceae family) (Chabner *et al.*, 2006; Shrivastava *et al.*, 2002). Camptothecin was isolated from the Chinese tree *Camptotheca acuminata* in 1966. Indian nathopodyte, *Foetida sluner* (Icacinaceae) (formerly known as *Nappia foetida*) has proved to be a rich source of camptothecin. The other valuble anticancerous agents, 9-methoxycamptothecin and 20-0-acetylcamptothecin have also been isolated from the same plant. Irinotecan and topotecan are currently the most widely used camptothecin analogues in the clinical cases of many cancers, colorectal, ovarian and small cell lung cancers. They interact with enzyme DNA topoisomerase I. Their binding to this nuclear enzyme allows single strand breaks in DNA and they damage DNA during replication; act in the S phase and arrest cell cycle at G_2 phase. Camptothecin is a specific inhibitor of DNA topoisomerase I. Inhibition of this enzyme is responsible for the antitumour activity. Camptothecin binds reversibly to a topoisomerase I DNA cleavable complex to form a stable complex. Camptothecin is a potent inhibitor of the growth of leukemia cells *in vitro* and shows remarkable antitumour activity against murine L1210 and P388 leukemia, and B16 melanocarcinoma *in vivo*. It is also given to cure liver carcinoma, and tumours of head and neck (Chabner *et al.*, 2006; Pandey and Madhuri, 2006a; Pandey *et al.*, 2013a).

Colchicum autumnale Linn. (Liliaceae plant family) is a rich source of colchicines, which has been used for a long time in the treatment of gout. During 1940 to 1950, the genetic potentialities and possible therapeutic value of this compound in combating cancer became apparent (Shrivastava *et al.*, 2002). The anticancer activity of colchicine is the result of its interaction with '*tubulin*', which aggregates to form the microtubules with mitotic spindle. Ellipticine-types of alkaloid have been isolated from the Apocyanacecae plant family members, *Ochrosia elliptica* and *Aspidosperma olivaceum*. Ellipticine and its two naturally-occurring analogues, 9-methoxyellipticine and olvacine have shown promising results as potential anticancerous drugs. Ellipticine-types of alkaloid significantly inhibit DNA, RNA and protein synthesis. The mode of action of these compounds against cancerous cells may be due to intercalation between the base pairs of DNA to bind to nucleic acid. *Jatropha gossypifolia* Linn. (Euphorbiaceae plant family) has been used ethnomedically for many years to treat cancer. The major active compound isolated from this is macrocyclic ditserpenoid, jatrophane with a new carbon skeleton and an unusual array of functionality. Bruceatin from *Brucea antidysentirca* plant is used in Ethiopia against cancer. Bruceatin acts through inhibition of protein synthesis. It exhibits a very high antileukemic activity. *Maytenus serrata* and its other species contain maytansine and ansa macrolide that are active against many neoplasms. Because of its high potency, small amount of maytansine is sufficient to treat many thousand patients. Another plant showing potent antileukemic properties is *Tripterygium wilfordii*, which has diterpenes and triptolide. This plant is potent antileukemic agent, but contains only small amounts and is not easily accessible (Pandey *et al.*, 2013a; Shrivastava *et al.*, 2002).

Taxol (a novel diterpenoid) is now one of the most promising anticancer agents. It was originally isolated from the bark of western/pacific yew tree (nut), *Taxus brevifolia* (Taxaceae family) (Chabner *et al.*, 2006; Shrivastava *et al.*, 2002). Presently, this compound is also isolated from other species of *Taxus*. Paclitaxel (taxol) and its congenic (semisynthetic docetaxel or taxotere) cause inhibition of mitosis and promote microtubule formation. Taxol enhances tubulin polymerization (a mechanism opposite to that of vinca alkaloids) and stabilizes microtubules, so the depolymerization microtubules are prevented. This stability results in inhibition of normal dynamic reorganization of the microtubule network that is essential for vital interphase and mitotic functions. It shifts monomer-polymer equilibrium of tubulin towards polymeric state. No significant effect of taxol on DNA, RNA and protein synthesis has been noted. Taxol has proved to be an exceptionally promising cancer chemotherapeutic agent with an unusually broad spectrum of potent antileukemic and tumour inhibiting properties. It is a diterpenoid compound that contains a complex taxane ring as its nucleus. The side chain linked to the taxane ring at carbon-13 is essential for its antitumour activity. Modification of the side chain has led to its identification of a more potent analog docetaxel (taxotere), which has clinical activity against breast and ovarian cancers. Taxol shows activity *in vivo* against P 388, P 1534 and L 1210 leukemias, B 16 melanocarcinoma, Lewis lung carcinoma, sarcoma 180 and CX-I colon, LX-I lung and MX-I breast xenographs. Taxol has also been proved to be active against the refractory advanced ovarian cancer and metastatic breast cancer. Thus, this drug has a central role in the combination therapy of cisplatin-refractory ovarian, breast, lung, oesophagus, bladder, and head and neck cancers (Chabner *et al.*, 2006; Pandey and Madhuri, 2006a; Pandey *et al.*, 2013a).

ProImmu (a polyherbal drug) contains four immunoactive plants, *viz.*, *Emblica officinalis* (Amla), *Ocimum sanctum* (Tulsi), *Tinospora cordifolia* (Giloe) and *Withania somnifera* (Ashwagandha). The mechanism of action of ProImmu is probably through the immunostimulation. It possesses better immunocompetence since it enhances the activity of cytotoxic T cells as well as NK cells. It potentiates the specific (humoral and cellular arms of the host immune system) and nonspecific immunity of the host. ProImmu increases the size, number and phagocytic activity of macrophages (*e.g.*, lymphocytes, monocytes, etc.). It also increases the lipopolysaccharide (LPS)-

induced proliferation of leucocyte (WBCs, *e.g.* lymphocytes and monocytes). A significant increase in the number of plaque forming cells (antibody producing cells) of spleen has also been noted in experimental animals. Thus, the stimulation of splenocytes to produce plaque forming cells by ProImmu helps in stimulating humoral arm of immunity. This drug has been scientifically proved to elicit the tumoricidal effect (Agrawala *et al.*, 2001; Nemmani *et al.*, 2002; Madhuri, 2008; Pandey and Madhuri, 2006a; Pandey *et al.*, 2013a).

    E. officinalis (Amla) is used in many diseases, and it is believed to increase the defense against diseases. It is particularly used for the treatment of cancer, diabetes, liver disorders, heart disease, ulcer, snake venom, haemorrhage, diarrhea, dysentery, anaemia and ophthalmic disorders. The antioxidant, immunomodulatory, anticancer, cytoprotective, analgesic, antimicrobial, antipyretic, antitussive and gastroprotective are the important properties of amla. Vitamin C, tannins and flavaniods present in amla have very powerful antioxidant activities. Due to rich in vitamin C, amla is successfully used in the treatment of human scurvy (Madhuri *et al.*, 2011b). *E. officinalis* has been reported (Sultana *et al.*, 2005) to elicit the immunostimulatory and antitumour effects (Jeena *et al.*, 2001). The fruit of *E. officinalis* contains 18 compounds that inhibit the growth of gastric and uterine cancer cells. Among these compounds, organic acid gallates together with hydrolysable tannins were found to be the major phenolic constituents (polyphenols) of *E. officinalis* fruit. The antiproliferative or antitumour activity of polyphenols might be linked to their antiinflammatory and antioxidant properties (Zhang *et al.*, 2004).

    A number of medicinal plants, traditionally used for thousands of years, are present in a group of herbal preparations of the Indian traditional health care system (Ayurveda) named '*Rasayana*', identified for their interesting antioxidant activities, indicating that *E. officinalis* also has antioxidant activity. *E. officinalis* is often incorporated in the herbal formulation called '*Triphla*', which contains the equal proportion of the fruits of *E. officinalis*, *Terminalia chebula* and *T. belerica*. Triphla, due to its strong antioxidant activity, restores the noise-stress induced changes. Albino rats were used to assess the immunomodulatory activities of Triphla on various neutrophil functions like adherence, phagocytic index, avidity index and nitro blue tetrazolium. Oral administration of Triphla appeared to stimulate the neutrophil functions in the immunized rats, and stress-induced suppression in the neutrophil functions was significantly prevented by Triphla. The presence of Triphla in diet had significantly lowered the B(a)P induced fore-stomach papillomagenesis in mice. Triphla also significantly increased the antioxidant status of animals which might have contributed to the chemoprevention. The cytotoxic effects of the AqE of Triphla were investigated on a transplantable mouse thymic lymphoma (barcl-95) and human breast cancer cell line (MCF-7). The differential response of normal cells and tumour cells to Triphala *in vitro* and the substantial regression of transplanted tumour in mice fed with Triphla have indicated its potential use as an anticancer drug for clinical treatment. Thus, this drug has been found to exhibit the chemopreventive potential (Khan, 2009; Madhuri *et al.*, 2011b; Pandey *et al.*, 2013a).

    The phenolic compounds, *viz.*, cirsilineol, cirsimaritin, isothymusin, apigenin and rosmarinic acid, and appreciable quantities of eugenol (a major component of volatile oil) obtained from the leaves and stems of *O. sanctum* possess good antioxidant activity. The leaves of *O. sanctum* suppressed the B(a)P induced chromosomal aberration in bone marrow and elevated the GSH and GST activities in mouse liver, suggesting the role of this herb against cancer. The alcoholic extract (AlE) of *O. sanctum* leaves also possess modulatory influence on the carcinogen metabolizing enzymes, such as aryl hydrocarbon hydroxylase and GST, which are important in detoxification of carcinogens and mutagens. The *O. sanctum* leaves blocked or suppressed the DMBA induced carcinogenesis by inhibiting the metabolic activation of this carcinogen. The *O. sanctum* leaf extract as well as their flavonoids orientin and vicenin have shown potent antioxidant and antilipid peroxidative activities that strongly suggest the *'free radical scavenging'* as a major mechanism by which *Ocimum* products protect against cellular damage and tumour. Conclusively, this herb possesses several biological and medicinal activities, including antioxidant, anticancer, chemopreventive and radioprotective activities (Madhuri, 2008; Pandey and Madhuri, 2010d; Pandey *et al.,* 2013a; Uma Devi, 2001).

The anticancer activity exhibited by *T. cordifolia* can be attributed to the presence of berberine alkaloid (Jagetia and Rao, 2006). The stem of *T. cordifolia* increases the humoral immune response during solid tumour growth. The immunomodulatory activity of the combined extracts of *O. sanctum, W. somnifera* and *E. officinalis* was noticed. In another study, the immunomodulatory effects of *W. somnifera* and *T. cordifolia* were observed. The methanolic extract (200 mg/kg, ip, daily for 5 days) of *T. cordifolia* stem to Balb/c mice has been found to increase the humoral immune response and reduced solid tumour growth (Madhuri, 2008; Madhuri *et al.,* 2011b; Pandey *et al.,* 2013a).

W. somnifera has been explained to elicit the antitumour effect. It scavenges reactive molecules, leading to antimutagenesis and anticarcinogenesis. The *in vivo* growth inhibitory effect of *W. somnifera* on transplantable mouse tumour, S180 has been observed. The anticancer effect of *W. somnifera* is probably due to action of its main constituent withanolides, *viz.,* withaferin A (which inhibits RNA and protein production) and withanolide D (which inhibits RNA production). The RNA and protein inhibition can lead to increased cancer cell death. The chemopreventive activity of *W. somnifera* root extract on induced skin cancer in mice was seen. This effect was due in part to the antioxidant/free radical scavenging activity. An *in vitro* study of *W. somnifera* showed that withanolides inhibited the growth in human breast, central nervous system (CNS), lung and colon cancer cell lines (Madhuri, 2008; Madhuri and Pandey, 2009c). Given its broad spectrum of cytotoxic and anticancer activity, *W. somnifera* presents itself as a novel therapy for cancer (Madhuri *et al.,* 2009c).

The phytotherapeutic agents, *e.g.,* curcumin, garlic, green tea, lentinan, quercetin, genistein, Asian ginseng and silymarin have been examined in ovarian, cervical, endometrial or vaginal cancer. Curcumin (isolated from *Curcuma* genus, including *Curcuma longa*) killed the ovarian cancer cells. Garlic (*Allium sativum*) significantly decreased the incidence of cervical carcinoma, indicating that it has preventive effect on cervical cancer. Green tea (*Camellia sinensis*) decreased the ovarian tumour weight. Quercetin is a flavonoid present in high concentrations in tea (*Camellia sinensis*), onion (*Allium cepa*), kale, French bean (*Phaseolus vulgaris*) and apple. It may synergize the chemotherapeutic agents. Genistein (an isoflavone, commonly isolated from the seeds of *Glycine max*) and quercetin showed synergistic antitumour activity against ovarian carcinoma. Lentinan (a polysaccharide, isolated from *Lentinus edodes*, the edible shiitake mushroom) can also inhibit endometrial tumour growth. Asian ginseng (*Panax ginseng*) is a traditional Chinese medicine having preventive effects on both cervical and ovarian carcinogenesis. Silymarin (a standardized extract of *Silybum marianum*) can inhibit the proliferation of ovarian cancer cells *in vitro*. It may also synergize some chemotherapeutic agents. Allicin, a major constituent of garlic inhibits the proliferation of human mammary, endometrial and colon cancer cells. Cucurbitacin E, a tetracyclic triterpenoid extracted from *Ecballium elaterium*, showed anticancer activity against human ovarian cancer cells (Madhuri, 2008; Pandey *et al.*, 2013a). The antiproliferation and apoptosis are induced by curcumin in human ovarian cancer. The chemopreventive activity of curcumin might be due to its ability to inhibit cell growth and induce apoptosis (Shi *et al.*, 2006). The anticancer effect of aloe-emodin (from the root and rhizome of *Rheum palmatum*) on human cervical cancer has been noted (Guo *et al.*, 2007).

In conclusion, this is most widely accepted by the scientists that the anticancer effect of almost all the plant-drugs or medicinal plants may be due to their immunostimulatory, phagocytic, antioxidant and various other tissue protective activities. The active principles of plant-drugs block the cancer cells in mitosis (*i.e.*, inhibition of mitosis), block the cancer cells at the S-G$_2$ interphase of cell cycle, inhibit the microtubule assembly, inhibit the DNA topoisomerase I and II nuclear enzymes, and inhibit the DNA, RNA and protein synthesis (Madhuri, 2008; Pandey *et al.*, 2013a).

Other Important Anticancer Medicinal Plants

Of course, many herbal products have been manufactured, characterized and found to possess significant chemotherapeutic action against various types of cancer, as illustrated above and cited ahead in different parts/chapters. Some of the important anticancer herbal drugs/medicinal plants are further described here.

Curative effect of primary stage of cancer is enhanced by combining the dietary supplementation (agricultural foods) of metabolic deficiency with a simultaneous specific stimulation of the immunity of patients. This combined biological, agricultural and immunological therapy has cured hundreds of cancer patients and experimental animals. Some dietary agricultural plants with anticancer properties include *Allium cepa* (Piyaz), *A. cepa* var. *aggregatum* (Shallot), *A. sativum* (Lasun), *Brassica campestris* (Sarson), *Br. oleracea* var. *botrytis* (Phoolgobhi), *Br. oleracea* var. *capitata* (Pattagobhi), *Br. rapa* (Shalgam), *Citrus limon* (Baranibu), *Curcuma longa* (Haldi), *Emblica officinalis* (Amla), *Glycine javanica* (Soybean), *Lycopersicon esculentum* (Tamatar), *Momordica charantia* (Karela), *Swertia chirata* (Chirayita), *Trigonella foenumgraecum* (Methi), *Triticum aestivum* (Gehun) and *Zingiber officinale* (Adrak). Some other food plants with anticancer activity are *Amorphophallus companulatus* (Suran), *Avena sativa* (oat), *Cajanus cajan* (Arhar), *Hordeum vulgare* (Jau), *Lens culinaris* (Masur), *Mentha arvensis* (Podina) and *Zea mays* (Makka) (Madhuri and Pandey, 2008).

Phytoestrogens are natural plant compounds which have a similarity to human estrogens. They can lower the risk of breast, prostate and colon cancer, and can reduce the incidence of heart disease and osteoporosis. They have the beneficial on the body's hormonal balance, acting as both agonists and antagonists. Phytoestrogens act as estrogen agonists by occupying ER sites when natural estrogens are unavailable. They also act as estrogen antagonists by occupying ER sites ahead of the body's natural estrogens and equally importantly ahead of synthetic estrogens, and also environmental estrogens derived from chemical products, otherwise known as *'bad estrogens or xenoestrogens'* (Pandey and Madhuri, 2011b).

The microbial antibiotics (chemotherapeutic agents) produced by fungi are very useful for the treatment of various cancers. Some of these antibiotics are actinomycin D (dactinomycin) isolated from *Streptomyces* sp.; bleomycin from *S. verticillus*; daunomycin from *S. coeruleorubidus*; doxorubicin, epirubicin and idarubicin from *S. peuceticus*; geldanamycin from *S. hygroscopicus;* and mitomycin C from *S. caespitosus*.

Actinomycin D is active against choriocarcinoma, Kaposi's sarcoma, Wilms' tumour, rhabdomyosarcoma and testis tumour. Bleomycin is used in cancers of germ-cell, cervix, head and neck. Daunomycin is given in acute myelogenous and acute lymphocytic leukemias. Doxorubicin is used for treatment of soft tissue, oesteogenic and other sarcomas; Hodgkins's disease; non-Hodgkins's lymphoma; acute leukemia; breast, genitourinary, thyroid, lung and stomach cancers; and neuroblastoma and other childhood sarcomas. Epirubicin is given in breast cancer and geldanamycin inhibits experimental cancer. Idarubicin is effective in breast cancer and leukemia. Mitomycin C is used for cancers of lung, stomach, colon, rectum and anus (Pandey and Madhuri, 2009b).

In a research study, the anticancer effect of ProImmu on EO induced ovarian adenocarcinoma was observed in rats. The EO (750 µg/kg, orally, weekly for 24 weeks) increased activities of enzymes, *viz.*, serum glutamate pyruvate transaminase (SGPT) and serum alkaline phosphatase (SAP) were significantly (P<0.05) decreased by ProImmu (500 mg/kg, orally, daily for 4, 8 and 12 weeks after 20, 16 and 12 weeks of EO administration in three groups of rats, respectively). The ovarian tissues of EO alone administered group revealed marked fibrous tissue proliferation of follicular epithelium. Various cancerous lesions, including hyperchormatosis, enlargement of nuclei, anisokaryosis, anisocytosis and proliferation of blood vessels filled with red blood cells (RBCs) were seen. The ProImmu treated (concominant with EO dosing) groups revealed mild to much more regeneration in the ovarian tissues, which indicated that ProImmu has anticancer effect on oestrogen induced ovarian cancer (Madhuri *et al.*, 2012b).

In another experiment, the anticancer activity of ProImmu against EO induced uterine cancer in rat was determined. Biochemically, EO (750 µg/kg, orally, weekly for 24 weeks) significantly (P<0.05) decreased the activities of glucose-6-phosphate dehydrogenase (G6PDH) and protein. However, ProImmu (500 mg/kg, orally, daily for 4, 8 and 12 weeks after 20, 16 and 12 weeks of EO dosing) significantly increased the activities of G6PDH and protein, and returned these parameters towards normal (Madhuri, 2008). Similarly, the normal activities of serum glutamate oxaloacetate transaminase (SGOT) and SGPT were significantly increased after administration of EO, which decreased significantly towards normal after treatment with ProImmu for 4, 8 and 12 weeks. Histopathologically, the uterine tissues damaged by EO revealed fibroblastic bundles made up of mature fibrocytes, which laid down as collagens. Focal hyperplasia of endometrial lining was quite evident. Other carcinogenic changes, including hyperchormasia, enlargement of nuclei, anisokaryosis, anisocytosis, new angiogenesis and glandular polarity were also noticed in the uterus. On the contrary, the uterine tissues of rats treated with ProImmu and EO both revealed mild to much less degree of changes, and increasing order of regeneration and normalization consequently after 4, 8 and 12 weeks of ProImmu treatment. This suggests that ProImmu has anticancer effect and, may treat the drug-induced cancers (Madhuri, 2008; Madhuri *et al.*, 2010).

The beneficial effects of ProImmu and its two ingredients, *viz.*, *T. cordifolia* and *W. somnifera* have been noted against EO altered haematological profiles in female albino rats. The EO (250 µg/kg, orally, thrice a week) altered the levels of haemoglobin (Hb), total leucocyte count (TLC) and differential leucocyte count (DLC, *viz.*, lymphocyte, monocyte, neutrophil and eosinophil) could be significantly normalized by ProImmu (150 mg/kg, orally, daily for 8 weeks); however, normalization of these parameters brought up by *T. cordifolia* (250 mg/kg, orally, daily for 8 weeks) and *W. somnifera* (250 mg/kg, orally, daily for 8 weeks) was found to be of lesser degree. ProImmu (50 mg/kg, orally, daily for 12 weeks), *T. cordifolia* (250 mg/kg, orally, daily for 12 weeks) and *W. somnifera* (250 mg/kg, orally, daily for 12 weeks) also caused the normalcy of these parameters, but to less extent than observed after 8 weeks (Madhuri *et al.*, 2012c).

The injectable formulation of centchroman showed antitumour activity against EAC in mice. The drug significantly (P<0.05) increased the median day of death and host life span. Lycovin syrup 1.5 ml and 7.5 ml per kg body weight administered orally, reduced the development of sarcoma induced by 20 MC by 35 per cent and 70 per cent, respectively. It also inhibited the hepatocarcinogenesis induced by NDEA. The anticarcinogenic activity of lycovin might be due to the inhibition of P-450 enzyme activity and subsequent inhibition of the production of ultimate carcinogen as well as scavenging of oxygen free radicals during promotion of the transformed cells. The 'Rasayanas' of *O. sanctum* and *W. somnifera* are useful to treat cancer. Rasayanas enhanced the NK cells activity in normal as well as in tumour bearing animals. *W. somnifera* rasayana also activated the macrophages. All the rasayanas stimulated the antibody dependent complement mediated tumour cell lysis (Pandey and Madhuri, 2006b).

Curcuma longa Linn. (Haldi, turmeric) belongs to the plant family *Zingiberaceae*. Its rhizome (root or Haldi) contains curcumin, zingiberine and curcuminoids. The maximum tolerated dose and LD_{50} of the 50 per cent ethanol extract of *C. longa* rhizomes was found to be 250 and 500 mg/kg, ip in rat, respectively (Dhar *et al.*, 1968; Pandey, 2011d). Rhizomes are stimulant, carminative, alterative, blood purifier, antiperiodic and tonic. They are also given in sprain, swelling, tumour and liver diseases (Chopra *et al.*, 2002; CSIR, 1986). The rhizomes are also effective in colon, bladder and prostate cancers, intravesical tumour, fibrosarcoma, HCC, oesophagal carcinogenesis, leukemia, stomach papilloma and solid tumours (Rao *et al.*, 2004). The pigment colour called *'curcumin'* of Haldi has shown antiinflammatory, antitumour and antioxidant properties. Evidences suggest that curcumin can suppress tumour initiation, promotion and metastasis. Pharmacologically, curcumin has been found to be safe and human clinical trials indicated no dose-limiting toxicity when administered at the doses up to 10 g per day (Pandey, 2011d). Curcumin (diferuloyl methane), the active principle of *C. longa* is documented with several medicinal properties. It is a well known anticancer agent, and is found to induce apoptosis. It is also a potent antioxidant and antiinflammatory agent. It showed the chemopreventive effect of curcumin against DENA/phenobarbital induced-hepatocarcinogenesis in Wister strain male albino rats, as pre- and co-treatment with curcumin for 14 weeks significantly prevented the biochemical alterations induced by DENA/phenobarbital (Jagadeesh *et al.*, 2009; Pandey, 2011d).

Chapter 4
Anticancer Plant-Drugs: Part I

Some Anticancer Drugs from Plant Sources

The cancer may be caused in one of the three ways: incorrect diet, genetic predisposition and via environment. At least 35 per cent of all cancers worldwide are caused by incorrect diet, and in the case of colon cancer, diet may account for 80 per cent of the cases. When one adds alcohol and cigarettes to diet, the percentage may increase to 60 per cent. The genetic predisposition gives rise to 20 per cent cancer cases. So, most cancers are associated with a host of environmental carcinogens. As long ago as 480 BC, Hippocrates called *'primary constitution of man'*, we today call *'genetics'*, and we can infer that foods *'resulting from human skills'* can be equated with today's diet (Pandey and Madhuri, 2008b). Now, more than 50 per cent of all modern drugs in clinical use are of natural products, many of which have the ability to include apoptosis in various cancer cells of human origin (Pandey, 2009; Pandey and Madhuri, 2008b; Rosangkima and Prasad, 2004).

Some Anticancer Plants

The anticancer activity of many plants has been reviewed earlier by the author and his associate (Madhuri and Pandey, 2008, 2009a and 2009b; Pandey, 2009 and 2011b; Pandey and Madhuri, 2006a, 2006b, 2008b, 2008c, 2009a, 2009b and 2011a). However, this topic includes the following anticancer medicinal plants/herbal drugs: *Abrus precatorius, Aglaia roxburghiana* (*A. elaeagnoidea* or *A. odoratissima*), *Allium sativum,* anacartin and serankottai nei (from *Semecarpus anacardium*), *Amorphophallus companulatus, Avena sativa, Brassica campestris, Br. oleracea* var. *botrytis, Br. oleracea* var. *capitata, Br. rapa, Cajanus cajan,* camptothecin and its analogs irinotecan and topotecan (from *Camptotheca acuminata*), *Cassia fistula,* centchroman, *Citrus limon, Crocus sativus, Curcuma longa, Emblica officinalis, Ervatamia heyneana* (*Tabernaemontana heyneana*), *Glycine javanica, Hippocratea murcantha, Hordeum vulgare, Hygrophila spinosa*

(*Asteracantha longifolia*), *Indigofera mysorensis*, magniferin (from *Canscora decussaca*), *Lens culinaris*, *Lycopersicon esculentum*, lycovin, *Mentha arvensis*, *Momordica charantia*, *Ocimum sanctum*, *Olea polygama*, paclitaxel (Taxol) and its analog docetaxel (Taxotere) (from *Taxus bravifolia*), plumbagin (form *Plumbago rosea or P. indica*), podophyllotoxins or epipodophyllotoxins etoposide and teniposide (from *Podophyllum peltatum* or *P. hexandrum*), ProImmu (a formulation of *E. officinalis*, *O. sanctum*, *Tinospora cordifolia* and *Withania somnifera*), 'Rasayanas' of *O. sanctum* and *W. somnifera*, *Solanum* sp., *Swertia chirata*, *Terminalia arjuna*, *Trigonella foenum-graecum*, *Triticum aestivum*, *Vanda parviflora*, vinca alkaloids, *viz.*, vinblastine, vincristine and vindesine (from *Catharanthus roseus*, *Vinca rosea* or *Lochnera rosea*), *Wedelia calendulacea* (*W. chinensis*), *W. somnifera*, *Zea mays*, *Zingiber capitatum* and *Z. officinale* (Pandey and Madhuri, 2008b; Pandey *et al.*, 2013a).

Henceforth, Table 2 (Pandey and Madhuri, 2008b; Pandey *et al.*, 2013a) represents some anticancer plants with their common names, main active components, mechanism of actions and cancers in which they are used. These anticancer agents have been found very effective in experimental and/or clinical cases of many cancers, *e.g.*, sarcoma, leukemia, lymphoma and carcinoma as reported by various workers. Many authors (Chopra *et al.*, 2002; CSIR, 1986; IDMA, 2002) have elucidated the main active components of these plants; however, their actions and uses against various cancers have been specifically reported by other authors (as mentioned under each anticancer plant).

Other Anticancer Plant-Drugs

Besides, certain microbe-derived anticancer agents (antibiotics) of plant origin are actinomycin from *Streptomyces* sp.; bleomycin from *S. verticillus*; daunomycin from *S. coeruleorubidus*; doxorubicin, epirubicin and idarubicin from *S. pneuceticus*; geldanamycin from *S. hygroscopicus*; and mitomycin C from *S. caespitosus* (Chabner *et al.*, 2006; Rang *et al.*, 2003). Actinomycin and daunomycin block DNA dependent RNA synthesis; bleomycin degrades preformed DNA; doxorubicin, epirubicin and idarubicin bind with DNA, and inhibit both DNA and RNA synthesis; geldanamycin causes cell cycle disruption; and mitomycin C after enzyme activation acts as bifunctional alkylating agent. Actinomycin is given in sarcoma and germ-cell tumours; bleomycin used in cancers of germ-cell, cervix and head and neck; daunomycin in leukemia (blood cancer); doxorubicin in lymphoma, sarcomas and cancers of breast, ovary and lung; epirubicin in breast cancer; geldanamycin inhibits experimental cancer; idarubicin in breast cancer and leukemia; and mitomycin C is given in gastric, colorectal, anal and lung cancers (Pandey and Madhuri, 2008b).

Some Ethnomedicinal Plants Useful in Cancer Remedy

It has been well recognized that allopathic drugs are not without danger as they exhibit severe toxicity on normal tissues (Pandey, 1990). Therefore, worldwide research is going on to investigate the best effective antitumour agents from different sources. Recent pharmacological researches revolve around the urgency to evolve suitable chemotherapeutic agents for the treatment of tumours (benign and malignant) without having toxic effects (Pandey and Madhuri, 2006a; Somkuwar, 2003). Medical

Table 2: Some Anticancer Plants

Botanical Name (with Common Name)	Active Principles	Action	Cancer in which Used (with References)
Acanthopanax gracilistylus	—	Antioxidant	Liver cancer cells (Lin and Huang, 2000)
Acer negundo	Saponins	—	Various cancers (CSIR, 1986)
Azadirachta indica (Neem)	Polyphenolic myoinositol, dexamethasone	Cytotoxic, antioxidant	Various cancers (Pandey *et al.*, 2013a), lung tumour of mice (Hecht *et al.*, 1999), liver and mammary cancers (Tepsuwan *et al.*, 2002)
Camellia sinensis (Tea)	Polyphenols, epigallocatechin-3-gallate	Apoptosis induction, cell cycle arrest	Tumours (Ahmad *et al.*, 1997)
Colchicum autumnale (Surinjan)	Colchicine 12, demecolcine 13	Antimitotic	Solid tumours and some forms of leukemia, especially in chronic myelocytic leukemia (Shrivastava *et al.*, 2002; CSIR, 1986).
Coriolus versicolor (Chinese herb)	Bis-benzyl-iso-alkaloids, bufalin, berberine, tetrandrine	Apoptosis induction, complexes with DNA	HL60 and U937 cells (Dong *et al.*, 1997)
Cylopia intermedia (Honeybush tea)	Polyphenolic compounds	Antioxidant, anti-mutagenic, alters P450-mediated metabolism	Various cancers (Marnewick *et al.*, 2000)
Eucalyptus grandis	Euglobal-G_1	—	Various cancers (Takasaki *et al.*, 2000)
Evodia officinalis	Quinolone alkaloids, flavonoid glucosides	Cytotoxic	Colon carcinoma (HT29), breast carcinoma (MCF7), hepato-blastoma (HepG2) (Xu *et al.*, 2006)
Fragaria vesca (Strawberry)	Vitamin C, bioflavonoids, chalcones	Antioxidant	Various cancers (Paiva and Russell, 1999)
Gymnosporia rothiana	GCE	DNA, RNA and protein synthesis inhibition by 12-36 hr	Leukemia in mice (Chapekar and Sahasrabudhe, 1981)
Ludwigia hyssopifolia	Piperine	—	Various cancers (Das *et al.*, 2007)
Malus domestica (Apple)	—	Antioxidant	Various cancers (Eberhardt *et al.*, 2000)
Olea europaea (Olive)	Polyphenols	Antioxidant	Various cancers (Langest, 1995)

Contd...

Table 2–Contd...

Botanical Name (with Common Name)	Active Principles	Action	Cancer in which Used (with References)
Ornithogalum sp.	Cholestane glycoside	Apoptosis induction	HL60 cells (Hirano et al., 1996)
Parquetina nigrescens	Ellipticine (pyridocarbazole alkaloid)	—	Tumours (Shrivastava et al., 2002)
Pinus pinaster (Maritime)	Polyphenolic fraction, biofla-vonoids, pro-anthocyanidins, procydin, pycnogenol	Antioxidant, increases NK cell activity, modulates mitogenic signals, induces apoptosis	DU145 cells, prostrate and skin cancers (Agarwal et al., 2000)
Punica granatum (Anar)	Alkaloids, antho-cyanidines, vitamin C	—	Solid tumour and ascites tumour in albino mice (CSIR, 1986; Pandey et al., 2013a)
Rhizoma zedoariae	β-elemene	Cell cycle arrest fromS to G_2M phase	(Zheng et al., 1997)
Ricinus communis (Arand)	Ricinine, triricinolein, ricinoleic acid	—	Tumours (Madhuri and Pandey, 2008)
Rubia cordifolia (Rosemary)	Carnosic acid, rosemary acid	DNA adducts formation	P388, L1210 and B16 melanoma cells (Poginsky et al., 1991)
Scutellaria radix, S. indica	Flavonoids	Prostaglandin E_2 production	Rat C6 glioma cells (Nakahata et al., 1998)
Trichosanthes kirilowi	Protein compounds (GLQ 223 and Q)	—	Tumours (Lakshmi Prabha et al., 2002)
Tripterygium wilfordii	Diterpenes, triptolide	—	Tumours (Shrivastava et al., 2002)
Uncaria tomentosa	—	Apoptosis induction	Tumour (Sheng et al., 1998)
Undaria pinnatifida (Seaweed)	Vivo-natural	Prophylactic	Lewis lung cancer in mice (Furusawa and Furusawa, 1985)
Vacciniumstamineum	—	Antioxidant, protein inhibitor, nuclear factor-kappa B	Human lung cancer, leukemia cells (Wang et al., 2007)
Vernonia guineensis	Vernolepin (sesquiterpene lactone)	—	Tumours (Shrivastava et al., 2002)

Contd...

Table 2–Contd...

Botanical Name (with Common Name)	Active Principles	Action	Cancer in which Used (with References)
Viscum album,Viscum var. *coloratum* (Korean mistletoe)	Lectin alkaloids	Caspase-3 activation, lectin 11-induced apoptosis, telomerase inhibition via mitochondrial pathway independent of p53	U937, HL60, lymphoblastoma and hepatocarcinoma cells (Duong Van Huyen *et al.*, 2001)
	Hexamethylene bioacetamide	Induction with telomerase, p53-dependent apoptosis	Human colon carcinoma, LoVo and leukemia cells (Zhang *et al.*, 2000)
Vitis rotundifolia (Muscadene berry)	Resveratrol	Antioxidant	Lung tumour in mice (Hecht *et al.*, 1999)

information referred in the old Indian literatures includes several medicinal herbs, which have been in the use for thousands of years, in one form or the other, under the indigenous system of medicine. The indigenous system of medicine (Traditional Indian Medicine) has several medicinal plants with versatile antitumour properties that need detailed research for the development of antitumour herbal drugs. The contribution of ethnomedicinal plants in discovering new drugs has been enormous for treating diseases like cancer, hypertension, diabetes, etc. (Madhuri and Pandey, 2008; Somkuwar, 2003).

Some medicinal plants and their products including vegetables, fruits and crops play an important role in the prevention from cancer. Many plant-derived products exhibit potent antitumour activity against several cancer cell lines (Madhuri and Pandey, 2008; Pandey, 2009; Polidori, 2003; Sivalokanathan *et al.*, 2005; Vecchia and Tawani, 1998). In view of this fact, certain ethnomedicinal plants having antitumour properties have been enumerated by Madhuri and Pandey (2009b).

Certain Antitumor Ethnomedicinal Plants

Since chemotherapy, radiation, etc. cause severe toxicity, herbal plants have become popular throughout the world nowadays, and are used as a therapy for tumours/cancers.

Many of these plants include *Abrus precatorious* (Ghungchi), *Aglaia roxburghiana* (Priyangu), *Cassia fistula*, *Catharanthus roseus* (*Vinca rosea*, Sadabahar), *Crocus sativus* (Saffron), *Ervatamia heyneana*, *Hygrophila spinosa* (Talmakhana), *Hippocratea murcantha*, *Indigofera mysorensis*, *Ocimum sanctum* (Tulsi), *Olea polygama*, *Plumbago rosea* (Chitra), *Podophyllum hexandrum*, *Semecarpus anacardium* (Bhela), *Solanum dulcamara*, *S. indicum* (Barhanta), *S. khasianum*, *S. surattense* (Kateli), *Terminalia arjuna* (Arjuna), *Trigonella foenumgraecum* (Methi), *Vanda parviflora*, *Wedelia calendulacea* (Pila Bhangra), *Withania somnifera* (Ashwagandha) and *Zingiber capitatum* (Madhuri and Pandey, 2009b).

Table 3 includes 62 ethnomedicinal plants that exhibit antitumour activity with other biological activities (Aruna and Sivaramkrishnan, 1990; Chopra *et al.*, 2002; CSIR, 1986; IDMA, 2002; Jain *et al.*, 2006; Jarald and Jarald, 2006; Kathiresan *et al.*, 2006; Kaushik and Dhiman, 1999; Madhuri and Pandey, 2008 and 2009b; Prajapati *et al.*, 2003; Rosangkima and Prasad, 2004; Vecchia and Tawani, 1998). Various combinations of active components of these plants, after isolation and identification can be made, and they have to be assessed for synergistic effects. Preparation of the standardized dose and dosage regimen may play a critical role in the therapy of tumours. Thus, there is a great need in searching for and manufacturing newer herbal drugs from the ethnomedicinal plants, which possess remarkable antitumour activity (Madhuri and Pandey, 2009b).

Certain Natural Products Against Cancer

Many *'natural products'* have been reported to act as chemoprotective agents against various types of cancer. These natural products are present in fruits, vegetables, plant extracts, herbs, microbes and marine organisms. A host of natural product constituents could be responsible for the protective effect against cancers, and it is

Table 3: Certain Antitumour Ethnomedicinal Plants

Botanical Name (with common name)	Family	Parts Used and their Main Active Components
Acanthus ilicifolius (Harcuch Kanta)	Acanthaceae	Whole plant contains resin, alkaloid, ribose derivatives of benzoxazoline
Adhatoda zeylanica (Arusa)	Acanthaceae	Whole plant contains pyrroloquinazoline alkaloid, vasicine, vesicinone, vesicinolone, hydroxyketone
Ajuga bracteosa (Nilkanthi)	Lameaceae	Whole plant contains glycoside, tannins, β and γ sitosterols, cerotic and palmitic acids
Alisma plantagoaquatica (Water plantain)	Alismataceae	Root and plant contain β sitosterol, lecithin, choline
Aloe vera (Ghee Kunwar)	Liliaceae	Leaf contains glycosides-anthracene derivatives: hydroxyanthraquinone derivatives
Alstonia macrophylla	Apocynaceae	Whole plant contains polyphenol, tannin, protein
Annona purpurea	Annonaceae	Stem and leaf contain alkaloid
Aphanmixis polystachya (Harinhara)	Meliaceae	Stem and seed contain limonoid, linoleic, stearic, oleic and palmitic acids, aphanamixin, aphananin
Apium graveolens (Ajmud)	Apiaceae	Whole plant contains isoquercitrin, choline, eudesmol alpha, β sedanenolide, oleoresin, limonene
Arnebia euchroma (Ratanjot)	Boraginaceae	Whole plant contains essential oils
Avena sativa (Oat, Jei)	Poaceae	Whole grain contains alkaloid, saponin, flavonoid, vitamin, protein
Bacopa monnieri (Brahmi)	Scrophulariaceae	Whole plant contains glycoside, saponin, triterpenoid, flavonoid, β sitosterol
Bauhinia variegata (Kachnar)	Caesalpiniaceae	Whole plant contains β sitosterol, kaempferol-3-glucoside, lupenol
Bruguiera gymnorhiza (Kankra)	Rhizophoraceae	Whole plant contains β sitosterol, lupenol, oleanolic acid
Butea monosperma (Dhak, Palas)	Fabaceae	Whole plant contains vitamin, protein, mineral
Cardiospermum halicacaburm (Kanphuti)	Sapindaceae	Whole plant contains alkaloid, β sitosterol, L-triacontanol, n-pentacosane, n-triacontane
Cleome viscose (Hulchul)	Capparideaceae	Whole plant contains macrocyclic diterpene, bicyclic diterpene, cleomeolide, coumarinoligan
Clerodendrum infortunatum (Bhates)	Verbenaceae	Root contains sterol, glycoside, campesterol, sitosterol, clerodin
Cylista scariosa	Fabaceae	Root contains tannin

Contd...

Table 3–Contd...

Botanical Name (with common name)	Family	Parts Used and their Main Active Components
Dianthus chinensis (Rainbow pink)	Caryophyllaceae	Whole plant contains eugenol, dianchinenosides (A, B, C and D)
Dregea volubilis (Nakchhikni)	Asclepiadaceae	Whole plant contains dregein glucoside
Eulophia nuda (Amarkand)	Orchidaceae	Tuber contains tabernaemontanine alkaloid
Excoecaria agallocha	Euphorbiaceae	Whole plant contains diterpene, linoleic acid
Glycyrrhiza glabra (Mulatti)	Fabaceae	Whole plant contains coumarin, saponin (glycyrrhizin and glabranin), isoflavone
Hagenia abyssinica	Fabaceae	Whole plant contains α and β kosins, kosotoxin, protokosin
Hordeum vulgare (Barley, Jau)	Poaceae	Whole plant contains amino acids (arginine, histidine, lysine, tyrosine and glycine)
Indigofera aspalathoides	Fabaceae	Whole plant
Kaempferia rotunda (Bhuichampa)	Zingiberaceae	Tuber contains essential oils
Lens culinaris (Masur)	Fabaceae	Seeds contain lipoidal, phytin, minerals, proteolytic enzyme
Leonurus sibiricus (Motherwort)	Lamiaceae	Whole plant contains alkaloids (leonurinine, leonuridine and leuronurine)
Lilium candidum	Lilliaceae	Bulb
Matthiola incana (Gilliflower)	Brassicaceae	Seed contains mucilage, fatty oil
Nicotina tabacum (Tobacco)	Solanaceae	Leaf contains nornicotine, piperidine, N- methylpyrroline, pyrrolidine, N-methyl-l-anatabine
Nothapodytes nimmoniana (Kalagaura)	Icacinaceae	Seed contains palmitic, oleic and linolenic acids
Operculina turpethum (Indian jalap, Nisoth)	Convolvulaceae	Root contains turpene, turpetheins (α and β)
Ophiorrhiza mungos (Sarahati)	Rubiaceae	Root contains β sistosterol, amorphous alkaloid
Oxalis acetosella (Rakta-Pushpa)	Oxalidaceae	Plant contains oxalic acid, potassium oxalate, vitamin C
Rhus succedanea (Kakrasingi)	Anacardiaceae	Leaf contains tannin, apigenin, glycoside
Rubia cordifolia (Manjit, Majith)	Rubiaceae	Root contains purpurin, pseudopurpurin, alizarin, xanthopurpurin (purpuroxanthin)
Rumex acetosa (Khatta Palak)	Polygonaceae	Leaf contains hyperoside, oxymethylanthraquinone, oxalic and tartaric acids
Salvadora persica (Jhak, Kharja)	Salvadoraceae	Leaf contains trimethylamine alkaloid, β sistosterol

Contd...

Table 3—Contd...

Botanical Name (with common name)	Family	Parts Used and their Main Active Components
Salvia officinalis (Sage)	Labiatae	Whole plant contains diterpene, phenolic acid, flavonoid, tannin
Solanum viride	Solanaceae	Leaf contains solanidine steroidal alkaloids
Stellaria semivestita	Caryophyllaceae	Herb
Stephania hernandiifolia (Nimuka)	Menispermaceae	Root, rhizome and plant contain d- and dl-tetrandrine, fangchinoline and d-isochondrodendrine alkaloids
Stereospermum personatum (Pader)	Bignoniaceae	Aerial part contains crystalline bitter substance
Stereospermumsuaveolens (Paral)	Bignoniaceae	Flower and whole plant contain bitter substance
Symplocos cochinchinecsis (Bholiya)	Symplocaceae	Bark contains alkaloid (loturine and colloturine)
Symplocos theaefolia	Symplocaceae	Leaf and twig
Syzygium cornocarpum	Myrtaceae	Whole plant contains oleanolic acid, eugenol
Terminalia catappa (Desi Badam)	Combretaceae	Whole plant contains oleic, palmitic and linoleic acids, flavonoid, tannin
Tinospora cordifolia (Amrita, Giloe)	Menispermaceae	Stem contains berberine, tinosporine, giloin
Tylophora indica (Jangli Pikvam)	Asclepiadaceae	Root and leaf contain alkaloid, α amyrin, quercetin
Uraria crinite (Dieng-kha-riu)	Fabaceae	Shrub
Urginea indica (Jangli Piyaz)	Lilliaceae	Bulb contains scillaren A (crystalline) and scillaren B (amorphous) glycosides
Viscum album (Ban, Banda)	Loranthaceae	Whole plant contains acetylcholine, proprionyl choline, lupeol, viscotoxin, flavonoid, sterol A
Vitex negundo (Nirgandi)	Verbenaceae	Leaf contains essential oil, nishindine alkaloid
Vitex trifolia (Amalbel)	Verbenaceae	Leaf contains camphene, diterpine, flavonoid
Xanthium strumarium (Chhota Gokhru)	Asteraceae	Root and whole plant contain sesquiterpene lactones (xanthinin, xanthinim and xanthatin)
Yucca aloifolia	Agavaceae	Flower contains aloifoline
Zea mays (Makka)	Poaceae	Whole plant contains β carotene, vitamins (C, E and K)
Zosima orientalis	Apiaceae	Root and fruit contain coumarin

likely that many of them play an important role. Although the mechanism of the protective effect is unclear, nevertheless, the consumption of fruits and vegetables lowers the incidence of cancer. A major group of these products are the powerful antioxidants, others are phenolic compounds and the remainder includes reactive groups that confer protective properties against cancers (Kaur and Kapoor, 2002; Madhuri and Pandey, 2008; Pandey, 2009; Polidori, 2003).

Some Anticancer Natural Products

There are a large number of natural products (Tables 4-7), which are derived from plants, microbes and marine organisms. They have been found very effective in experimental as well as clinical cases of many cancers, *e.g.*, sarcoma, lymphoma, carcinoma and leukemia. The plant-derived anticancer agents (*e.g.*, docetaxel, flavopiridol, irinotecan, paclitaxel, topotecan, vinblastine and vincristine; Table 5) and the microbe-derived anticancer agents (*e.g.*, antibiotics of anthracycline, bleomycin, actinomycin, mitomycin and aureolic acid groups; Table 6) have different mechanism of action with high anticancer properties. However, the marine-derived anticancer agents (*e.g.*, aplidine, bryostatin, cryptophycin, discodermolide, dolastatin, ecteinascidin and halicondrin B) have been shown to inhibit the experimentally induced cancers (Pandey, 2009; Pandey *et al.*, 2013a).

Treatment of Neoplasms by Certain Plants

The isolation and purification of potent active principles of plants involved in the treatment of neoplasm are essential. There is a great need for searching and developing newer medicinal plants possessing increased bioactivity and antineoplastic activity with their mechanism of action. Of course, the current approaches are being carried frequently in pharmacological research in evolving herbal drugs, which are endowed with potential interference with malignancy of cancer cells. Consequently, a number of antineoplastic herbal drugs (*e.g.*, vinblastine, vincristine, vindesine, ProImmu, etoposide, teniposide, anacartin, Serankottai nei, docetaxel, irinotecan, topotecan, etc.) have been marketed in recent past to be employed for the treatment of neoplasms/cancers (Pandey and Madhuri, 2006a; Pandey *et al.*, 2013a). In this context, some of the medicinal plants/herbal drugs are discussed here.

Some Plant-Drugs for Treatment of Neoplasm

Seeds of *Abrus precatorious* Linn. (Ghungchi) yield abrin, a galactose specific lectin. Abrin has been found to suppress the Ehrlich ascites carcinoma (EAC) growth in mice. The protein extract of *A. precatorious* seeds exhibited antitumour activity on Yoshida sarcoma in rats and mice (CSIR, 1986). Antitumour effect of abrin (7.5 mg/kg, every alternate day for 10 days) was reported on transplanted tumour in mice, and abrin was effective in reducing the solid tumour mass development induced by Dalton's Lymphoma Ascites (DLA) and EAC cells. An increased life span of ascites tumour bearing mice was also noticed after intraperitoneal administration of abrin (Ramnath *et al.*, 2002).

Table 4: Some Anticancer Plants

Botanical Name (with common name)	Main Active Components	Mechanism of Action	Cancer in which Used
Acanthopanax gracilistylus	—	Antioxidant	Liver cancer
Allium cepa Linn. (Piyaz), A. porrum (Leek)	Diallyl disulphide, quercetin flavonoid, allicin, allin	Detoxifies carcinogen, inhibits Helicobacter pylori, arrests cell cycle from S to G_2M phase	Cancers of lung and other organs, stomach cancer
Azadirachta indica Juss. (Neem)	Polyphenolic myoinositol, dexamethasone	Cytotoxic, antioxidant	Various cancers, lung tumour and liver cancer
Camellia sinensis (Green tea, black tea)	Polyphenols	Apoptosis induction, cell cycle arrest	Various tumours
Coriolus versicolor (Chinese herb)	Bisbenzyl isoalkaloids, berberine, bufalin, tetrandrine	Apoptosis induction, complexes with DNA	HL60 and U937 cancer cells
Cylopia intermedia (Honey-bush tea)	Polyphenolic compounds	Antioxidant, antimutagenic, alters P450-mediated metabolism	Various cancers
	Valepotriates	Cytotoxic	Various cancers
Emblica officinalis Gaertn. (Amla)	Ascorbic and phyllembic acids	Antioxidant, antitumour, immuno-modulatory	Breast, gastric and uterine cancers[1]
Eucalyptus grandis	Euglobal-G_1	—	Various cancers
Fragaria vesca Linn. (Strawberry)	Vitamin C, bioflavonoids, chalcones	Antioxidant	Various cancers
Gymnosporia rothiana	GCE	DNA/RNA and protein synthesis inhibited after treatment for 12-36 hr	Leukemia in mice
Malus domestica (Apple)	—	Antioxidant	Various cancers
Ocimum sanctum Linn. (Tulsi)	Flavonoids (orientin, vicenin, eugenol), oleanolic acid	Antioxidant, antitumour, immuno-modulatory	Various cancers
Olea europaea Linn. (Olive)	Polyphenols	Antioxidant	Various cancers
Ornithogalum sp.	Cholestane glycoside	Apoptosis induction	HL60 cells

Contd...

Table 4—Contd...

Botanical Name (with common name)	Main Active Components	Mechanism of Action	Cancer in which Used
Pinus pinaster (Maritime)	Polyphenolic fraction, ferrulic acid, bioflavonoids, proanthocyanidins, procydin, pycnogenol	Antioxidant, increases activity of NK cells, modulates mitogenic signaling, induction of G_1 arrest and apoptosis	DU145 cells, prostrate and skin cancers
Rhizoma zedoariae	β elemene	Cell cycle arrest from S to G_2M phase	Various cancers
Ricinus communis Linn. (Arand)	Ricinine and ricinoleic acid	—	Various tumours
Rubia cordifolia (Rosemary)	Carnosic acid, rosemary acid RC-1	Forms DNA adducts	P388 cells, L1210 cells, B16 melanoma
Scutellaria radix, *S. indica*	Flavonoids	Prostaglandin E_2 production	Rat C6 glioma cells
Solanum sp.	Flavonoid (quercetin), alkaloids (solasodine, solanine, solamargine)	—	Various tumours
Swertia chirata Buch.-Ham. (Chirayita)	Swertianin, swertinin, chiratanin and swertenol	—	Various tumours
Triticum aestivum Linn. (Wheat)	Ellagic, linolenic and oleic acids, sterols, phytases, vitamin E	Protects against lipid peroxidation	Skin cancer
Uncaria tomentosa	—	Apoptosis induction	Various tumours
Undaria pinnantifida (Seaweed)	Vivo-natural	Prophylactic	Lewis lung cancer in mice
Viscum album, *Viscum* var. *coloratum* (Korean mistletoe)	Lectin alkaloids	Caspase-3 activation, lectin 11-induced apoptosis, inhibition of telomerase via mitochon-drial controlled pathway independent of p53	U937, HL60, lymphoblastoid and hepatocarcinoma cells
	Hexamethylene bioacetamide	p53-dependent apoptosis, induction with telomerase	Human colon carcinoma, leukemia
Vitis rotundifolia (Muscadene berry)	Resveratrol	Antioxidant	Lung tumour in A/J mice
Withania somnifera Dunal (Ashwagandha)	Withanolides (withaferin A, withanolide D)	Antioxidant, antitumour, immuno-modulatory	Various cancers

Table 5: Plant-Derived Anticancer Agents

Compound	Plant Source	Mechanism of Action	Cancer in which Used
Docetaxel	Western yew tree (Taxanes)	Promotes tubulin assembly, inhibits microtubule deploymerization, also acts as a mitotic spindle poison, induces mitotic block in proliferative cells	Breast, ovarian, lung, head, neck and colorectal melanomas
Flavopiridol	*Dyboxylum binectiferum*	CDK modulator	Various cancers
Irinotecan	*Camptotheca acuminata*	Inhibits action of topoisomerase I, prevents religation of DNA strand, causes cell death	Leukemia, liver, colorectal, head and neck cancers
Paclitaxel	Western yew tree (Taxanes)	Promotes assembly of microtubules, stabilizes them against depolymerization, inhibits cell replication, causes apoptosis	Breast and ovarian adenocarcinomas, other solid tumours
Topotecan	*Camptotheca acuminata*	Inhibits topoisomerase I, repairs nuclear DNA	Ovary and lung cancers
Vinblastine	*Catharanthus roseus* (*V. rosea*)	Inhibits microtubule formation, arrests mitosis in metaphase	Breast, lymph, germ cell and renal cancers
Vincristine	*Catharanthus roseus* (*V. rosea*)	Inhibits microtubule formation, arrests mitosis in metaphase	Leukemia, lymphoma, breast and lung cancers

Table 6: Microbe-Derived Anticancer Agents

Compound	Microbial Sources	Mechanism of Action	Cancer in which Used
Actinomycin	Streptomyces sp.	Blocks DNA dependent RNA synthesis	Sarcoma, germ cell tumours
Bleomycin	Streptomyces verticillus	Metal chelating glycopeptide antibiotic, degrades preformed DNA	Germ cell, cervix, head and neck cancers
Daunomycin	Streptomyces coeruleorubidus	Blocks DNA dependent RNA synthesis	Leukemia
Doxorubicin	Streptomyces pneuceticus	Binds with DNA, inhibits DNA and RNA synthesis	Cancers of breast, ovary, lung and other organs
Epirubicin	Streptomyces pneuceticus	Binds with DNA, inhibits DNA and RNA synthesis	Breast cancer
Geldanamycin	Streptomyces hygroscopicus	Cell cycle disruption	Experimental cancers
Idarubicin	Streptomyces pneuceticus	Binds with DNA, inhibits DNA and RNA synthesis	Breast cancer, leukemia
Mitomycin C	Streptomycescaespitosus	Acts as bifunctional alkylating agent after enzyme activation	Gastric, colorectal, anal and lung cancers
Rapamicin	Streptomyces hygroscopicus	Immunosuppressant	Experimental
Streptozocin	Streptomyces achromogenes	—	Gastric and endocrine tumours
Wortmannin	Talaromyces wortmanni	Potent enzyme inhibitor	Experimental cancers

Table 7: Marine Organism-Derived Anticancer Agents

Compound	Mechanism of Action and use in Various Cancer
Aplidine	Experimental inhibition of cell cycle progression
Bryostatin 1	Experimental activation of protein kinase C (PKC)
Citarabine	Experimental inhibition of DNA synthesis used in leukemia and lymphoma
Cryptophycin	Experimental inhibition hyperphosphorylation of B-cell lymphoma
Discodermolide	Experimental stabilization of tubulin
Dolastatin 10	Experimental inhibition of microtubules and pro-apoptotic effects
Ecteinascidin 743	Experimental alkylation of DNA
Halicondrin B	Experimental interaction with tubulin

The whole plant extract of *Aglaia roxburghiana* Miq./Hiern in part, non Miq. (*A. elaeagnoidea* Benth., *A. odoratissima* Blume, Priyangu) and *Zingiber capitatum* have been found to protect the rats by stimulating the rejection of malignant growth, leading to healing. The extracts of *Cassia fistula* Linn. (stem), *Hippocratea murcantha* (aerial part) and *Indigofera mysorensis* (aerial part) decreased the rate of tumour growth and converted the malignant tumours into benign fibrous masses. However, the extracts of *Olea polygama* (aerial part) and *Vanda parviflora* Lindl. (whole plant) prolonged the log-phase of malignant tumour growth to 12 days by decreasing the growth rate, and the malignant tumour growth converted into benign hard fibrous mass (Babbar *et al.,* 1979).

Magniferin, a glycoside, isolated from *Canscora decussaca* Schult. (Sankhaphuli) has been reported to provide marginal immunity against L1210, P388, sarcoma 180 (S180), fibrosarcoma and EAC (Bhattacharya *et al.,* 1976). *Crocus sativus* Linn. (saffron) has been reported (Nair *et al.,* 1991) to act as an anticancer agent. The dried stigmas and tops of the styles of *C. sativus* constitute the saffron, which contains crocin (a yellow glycoside). The antitumour activity of saffron extract (200 mg/kg, orally) was observed against intraperitoneally transplanted S180, EAC and DLA tumours in mice, as a result of which the life span of respective tumour bearing mice was increased to 111.0, 83.5 and 112.5 per cent. The same extract was cytotoxic to P388, S180, EAC and DLA tumour cells *in vitro*. Thymidine uptake studies indicated the mechanism of action of saffron extract at the site of DNA synthesis.

The anticancer activity of root, stem and leaf extracts of *Ervatamia heyneana* Cooke (*Tabernaemontana heyneana* Wall.) was achieved against P388 and L1210 lymphocytic leukemias (Chitnis *et al.,* 1971). The processed extract of *Hygrophila spinosa* T. Anders. (*Asteracantha longifolia* Nees, Talmakhana) has shown the antitumour activity with increase of life span by 60.6 and 44.4 per cent in EAC treated mice, and 49.9 and 31.6 per cent in S180 treated mice (Mazumdar *et al.,* 1997). The tumour inhibiting and radiomodifying effects of plumbagin, a naphthoquione, isolated from *Plumbago rosea* Linn. (*P. indica* Linn., or Chitrak, Lal Chitra in Hindi) were studied on mouse EAC. Its acute LD_{50} in normal mice was found to be 9.4 mg per kg body weight. Plumbagin (@ 2-6 mg/kg, ip as a single dose) caused the inhibition of exponentially growing tumours. Multiple dose treatment of plumbagin, starting from 24 hr after tumour cell inoculation (a total dose of 9 mg/kg administrated in three fractions of 3 mg/kg, once daily) caused the maximum per cent increase in life span and tumour free survival. The combination of radiation (radiation therapy, 7.5 Gy to the abdomen) after the first plumbagin dose (1-3 mg/kg/fraction) synergistically increased the mouse survival at 120 days. Tumour inhibitory effect was less pronounced when treatment was started at more advanced tumour stages, but the combination of low dose fractions (2.5 or 3 mg/kg/fraction) with radiation therapy enhanced the per cent increase in life span and animal survival. Higher dose fraction in combination with radiation was not tolerated by mice. DNA appeared to be the likely target of plumbagin cytotoxicity (Kini *et al.*, 1997).

ProImmu (a polyherbal drug, containing the extracts of *E. officinalis, O. sanctum, T. cordifolia* and *W. somnifera*) is recommended in an adult human at the rate of one to two capsules, once or twice daily for 10 days or more. It possesses better immunocompetence when administered in normal as well as in immunocompromised individuals. It is also indicated against lymphopenia in cancer patients, and may restore the normal histoarchitecture of the body cells. The enhanced activity of cytotoxic T cells as well as NK cells by Immu-21 (ProImmu) indicates its usefulness as an adjuvant in viral and tumour chemotherapy. Immu-21 (30 mg/kg, ip for 14 days and 1 mg/kg for 21 days) significantly increased the lipopolysaccharide-induced leucocyte proliferation. The NK cell activity was significantly increased in mice pretreated with Immu-21 (10 and 30 mg/kg, ip, once a day for 7 days), while proliferation of splenic leucocyte to B cell mitogen, lipopolysaccharide and cytotoxic activities against K562 cells were selectively increased. Immu-21 showed significant increase in size, number and phagocytic activity of macrophages, leading to tumoricidal activity. Immu-21 induced increase in macrophage number may be due to stimulation of bone marrow stem cells, leading to their maturation towards monocyte-macrophage. ProImmu (Immu-21) has been stated to be useful against cancer as it maintains antibody titres, improves lymphocytes level, and prevents leucopenia, genotoxicity/mutagenicity and bone marrow suppression. Immu-21 (50 mg/kg, orally, daily for 20 days) significantly potentiated the humoral immunity in rabbits, and showed significant protection against UV rays, cyclophosphamide and cyclosporine A induced immunosuppression. A significant increase in the number of plaque forming cells (antibody producing cells) of spleen was noticed in experimental animals. The immunostimulatory effect of Immu-21 is dose-dependent. Stimulation of splenocytes to produce plaque forming cells by Immu-21 helps in stimulating humoral arm of immunity in the hosts. Beside anticancer activity, ProImmu is supportive to antibiotic therapy, and is indicated in major surgery, burn injury, multiple trauma and lymphopaenic cancer, to improve immunocompetence, in infertility due to toxoplasmosis, and for optimizing vaccinial response, etc. Immu-21 is safe and non-toxic as it did not cause any mortality in albino rats, following a single oral dose of 5 g per kg. It has a wide range of medicinal values. Chronic toxicity study of Immu-21 up to the dose of 400 mg per kg as against the recommended dose of 20 mg per kg daily for 90 days in rats revealed that it was non-lethal, non-toxic to kidney and liver, as well as free from haemopoietic toxicity. There was also improvement in liver function tests (LFTs) as evident by decreased bilirubin, SGOT and SGPT levels

(Agrawala *et al.*, 2001; Madhuri, 2008; Nemmani *et al.*, 2002; Pandey and Madhuri, 2006b; Pandey *et al.*, 2013a).

Semecarpus anacardium Linn. f., commonly called *'marking nut'* has wide use against various tumours. A single injection of SAN-AB (chloroform extract of the whole marking nut) could bring about complete inhibition of Yoshida sarcoma ascites tumour growth in rats. Anacartin forte, an Ayurvedic preparation of marking nut exhibited a broad spectrum anticancer activity and showed very satisfactory results in the cancer of oesophagus, liver and urinary bladder, and chronic leukemia. This preparation has selective action, attacking only the cancer cells without harming the normal cells. The antitumour activity of the chloroform extract of *S. anacardium* caused increase in life span during B16 melanoma, glioma 26, and L1210, P388 and advanced P388 leukemia cancers. *In vitro* effects of acetyl derivatives from *S. anacardium* nut indicated that the incorporation of radio-labeled precursors into DNA, RNA and protein was considerably inhibited at a concentration ranging from 40 to 75 μg per ml within 2 hr, and thereby biosynthesis of DNA, RNA and protein was significantly inhibited. Furthermore, Serankottai nei (from the purified nut milk extract of *S. anacardium*) possessed protective effect against aflatoxin B_1 (AFB_1) induced HCC. Serankottai nei (200 mg/kg/day for 14 days by oral gavage) restored back the normalcy in tumour bearing rats. The fruit extract of *S. anacardium* is effective against human epidermoid carcinoma of nasopharynx in tissue culture. *S. anacardium* nut has now been proved remarkably as an anticancer, antioxidant, membrane stabilizing and immunomodulatory agent (Pandey and Madhuri, 2006a; Pandey *et al.*, 2013a; Premalatha and Rajgopal, 2005).

Some species of *Solanum* plant possess antitumour activity. The AlE of dried rhizomes and roots, also that of flowering and fruiting twigs of *S. dulcamara* Linn., and extract of *S. surattense* Burm. f. (*S. xanthocarpum* Schard. and Wendl., Kateli) exhibited significant tumour-inhibiting activity against S180 in mice. The extracts of *S. indicum* Linn. (Barhanta) affect human epidermal carcinoma of the nasopharynx in tissue culture and on Friend-virus leukemia (solid) in mice. *S. dulcamara* and *S. indicum* both contain solanine and solanidine alkaloids, besides others as active principles; however, *S. surattense* contains solanocarpine, solanine-S and solanidine-S glucosidal alkaloids as active principles. *S. khasianum* C.B. Clarke emend. has also been reported to show anticancer activity. A steroidal glycoside, solasodine from the fruits of *S. khasianum* was isolated and the cell growth inhibition study of solasodine hydrochloride was performed. This agent showed greater percentage (up to 94.96 per cent) of cell growth inhibition than the standard anticancer drug in human myeloid leukemia cells line (U937) (Pandey and Madhuri, 2006a; Pandey *et al.*, 2013a).

The effect of ethanolic extract of *Terminalia arjuna* Roxb. W. and A. (Arjuna) bark on carbohydrate metabolizing enzymes of DEN induced HCC in Wistar albino rats was observed. Some plasma and liver glycolytic enzymes were significantly increased in cancer induced animals; while glyconeogenic enzyme, *i.e.*, glucose-6-phosphatase was decreased. These enzymes were reverted significantly to near normal range in treated animals after oral administration of *T. arjuna* for 28 days. The modulation of the enzymes constitutes the depletion of energy metabolism leads to inhibition of cancer growth. This inhibitory activity may be due to the anticancer activity of constituents present in the ethanolic extract of *T. arjuna* (Silvalokanathan *et al.,* 2005).

Trigonella foenum-graecum Linn. (Fenugreek) seed extract has been reported to show the antineoplastic activity against EAC model in balb C mice (Sur *et al.,* 2001). The ethanolic extract of *Wedelia calendulacea* Less., non Rich. (*W. chinensis* Merrill, Pila Bhangra) inhibited the growth of EAC (CSIR, 1986). The alcoholic extract (500 mg/ kg, orally) of *W. calendulacea* showed hepatogenic potency by producing hepatic regeneration in paracetamol induced liver damage in mice. This herb prevents and repairs the liver damage presumably by stimulating the hepatic microsomal drugs metabolizing enzymes and mitochondrial enzymes. Thus, this herb may be used in a variety of liver disorders including hepatomegaly, hepatic encephalopathy and HCC (Pandey, 1990; Pandey and Madhuri, 2006a; Pandey *et al.,* 2013a).

Medicinal Plants for Treatment of Cancer

A large number of chemopreventive agents are used to cure cancers, but they produce side effects that prevent their extensive usage. Although more than 1500 anticancer drugs are in active development with over 500 of the drugs under clinical trials, there is an urgent need to develop much effective and less toxic drugs (Kathiresan *et al.,* 2006). Medicinal plants have been stated (Jain *et al.,* 2006) to comprise about 8000 species and account for approximately 50 per cent of all the higher flowering plant species of India. In other words, there are about 400 families of the flowering plants; at least 315 are represented by India. Medicinal properties of few such plants have been reported but a good number of plants still used by local folklore are yet to be explored. Ayurveda, Siddha and Unani systems of medicine provide good base for scientific exploration of medicinally important molecules from nature. The rediscovery of Ayurveda is a sense of redefining it is modern medicines. Emerging concept of combining Ayurveda with advanced drug discovery programme is globally acceptable. Traditional medicine has a long history of serving peoples all over the world. The ethnobotany provides a rich resource for natural drug research and development. In recent years, the use of traditional medicine information on plant research has again received considerable interest. The Western use of such information has also come under increasing scrutiny and the national and indigenous rights on these resources have become acknowledged by most academic and industrial researchers (Garg *et al.,* 2007).

Preparation of standardized dose and dosage regimen may play a critical role in the phytoremedy of cancer. The rate with which the cancer is progressing, it seems to have an urgent and effective effort for making good health of humans as well as animals. Henceforth, there is a broad scope to derive the potent anticancer agents from medicinal plants, which need thorough research (Pandey and Madhuri, 2009a; Pandey *et al.,* 2013a).

Some Anticancer Plants

Here, several medicinal plants (Tables 8 and 9) with anticancer activities have been illustrated. Many of these medicinal plants contain various kinds of antioxidants. Hence, different combinations of the active components of these plants after isolation

Table 8: Some Anticancer Plants

Botanical Name (with Hindi/common Name)	Family	Main Active Components	Parts Used
Acrorus calamus (Bach)	Araceae	Asarone, eugenol, methyl eugenol, palmitic acid, camphene	Rhizome
Agrimonia pilosa (Hairy agrimony)	Rosaceae	Agrimonolide, flavonoid, tannin, triterpene, coumarin	Whole plant
Alphitonia zizphoides	Rhamnaceae	Zizphoisides (A, C, D and E triterpenoid saponins)	Whole plant
Alstonia scholaris (Devil tree)	Apocynasaceae	Triterpene, latex	Bark
Amorphophallus companulatus (Suran)	Araceae	Leucine, isoleucine, lysine stigmasterol, β sitosterol	Corm
Andrographis paniculata (Kalmegh)	Acanthaceae	Flavonoid, andrographin, andrographolide.	Whole plant
Avicennia alba	Avicenniaceae	Napthoquinolines and their analogues (avicequinones A, B and C)	Whole plant
Azadirachta indica (Neem)	Meliaceae	Tannin, β sitosterol, nimbin, quercetin, carotene	Bark, leaf, flower
Bruguiera exaristata	Rhizophoraceae	Alkaloid, inositol	Whole plant
Bruguiera paviflora	Rhizophoraceae	Tannin, phenolic compounds	Whole plant
Caesalpinia bonduc (Kantkarej)	Caesalpiniaceae	Caesalpins (α, β, γ, δ and e types), homoisoflavone	Whole plant
Cajanus cajan (Arhar)	Fabaceae	Many essential amino acids	Leaf, seed
Calophyllum inophyllum (Sultanachampa)	Clusiaceae	Quercetin, xanthone, biflavonoid, neoflavonoid, benzophenone, β sitosterol	Whole plant
Camellia sinensis (Green tea, black tea)	Theaceae	Polyphenols, epigallo-catechin-3-gallate, carotene, ascorbic acid, xanthine, inositol	Leaf
Cassia absus (Chaksu)	Caesalpiniaceae	Chrysophanol, isochrysophanol, rhein, β sitosterol	Leaf
Cayratia carnosa (Amalbel)	Vitaceae	Hydrocyanic acid, delphinidin cyaniding	Whole plant
Ceiba pentandra (Safed Simal)	Bombacaceae	Sesquiterpene lactone, lignin	Root, bark
Cissus quadrangularis (Hadjod)	Vitaceae	Tetracyclic triterpenoid, β sitosterol	Whole plant
Citrus limon (Nibu)	Rutaceae	Flavonoid, flavone, limonoid, limonene, nobiletin, tangeretin	Fruit
Cycas rumphii (Kama)	Cycadaceae	Resin	Bud, flower

Contd...

Table 8–Contd...

Botanical Name (with Hindi/common Name)	Family	Main Active Components	Parts Used
Decaspermum fructicosum (Christmas bush)	Myrtaceae	Plant contains essential oil, coumarins (ellagic acid derivatives)	Whole plant
Equisetum hyemale (Common horsetail)	Equisetaceae	Dimethlsulfone, kaempferol-diglucoside, caffeic acid	Whole plant
Eugenia caryophyllata (Laung, clove)	Myrtaceae	Volatile oils (eugenol, actyl eugenol and pinene), tannin	Whole plant, flower bud
Geranium robertianum (Herb Robert)	Geraniaceae	Geranin, tannin, citric acid	Whole plant
Glycyrrhiza glabra (Mulathi)	Fabaceae	Triterpenoid saponin (glycyrrhizin and glabranin), isoflavone, coumarin, triterpene (β-amerin stigmasterol), eugenol, indole	Rhizome
Ipomoea batatas (Sakkarkand)	Convolvulaceae	Monophenolase, catalase, cytochrome c-oxidase, anthocyanins, caffeic acid	Stem (tuber)
Mallotus philippensis (Sindur, Kamala)	Euphorbiaceae	Kamlolenic, conjugated dienoic, oleic, lauric, palmitic and stearic acids	Whole plant
Mentha arvensis (Podina)	Lamiaceae	Essential oils (menthol, menthone and limonene)	Whole plant
Moringa oleifera (Mungana)	Moringaceae	Vitamins A and C	Leaf, Root
Mussaenda raiateenisis	Rubiaceae	Quercetin, β sitosterol, saponin, glucoside	Bark
Pandanus odoratissimus (Kevda)	Pandanaceae	Dipentene, d-linalool	Whole plant, leaf
Pastinaca sativa (Parsnip)	Umbelliferae	Plant contains essential oil, crystalline furocoumarin	Whole plant
Physalis angulata (Wild tomato)	Solanaceae	Selenium, ayanin (flavonoid), β sitosterol	Whole plant, leaf
Piper longum (Pipli)	Piperaceae	Monocyclic sesquiterpene	Whole plant
Pongamia pinnata (Karanj)	Fabaceae	Dikitonepongamol, glabrin, karanjin	Root, fruit
Premna obtusifolia (Agetha)	Verbenaceae	Alkaloids (premnine, ganiarine and ganikarine)	Whole plant
Taxodium distichum	Taxaceae	Taxol (diterpene)	Seed
Tetragonia tetragonioides	Tetragoniaceae	Ca, Fe, vitamins A, B and C	Whole plant
Thespesia populnea (Paras-Papal)	Malvaceae	Glycosides of quercetin, isoquercitrin, kaempferol 3-flucoside, lupenone, β sitosterol	Stem
Vernonia cinerea (Sahadeyi)	Asteraceae	Lupeol, stigmasterol, β sitosterol	Whole plant

Table 9: Certain other Anticancer Plants

Botanical Name	Family	Parts Used
Allium bakeri	Liliaceae	Bulb
Berberis aristata	Berberidaceae	Whole plant
Cedrus deodara	Pinaceae	Seed
Celitis africana	Ulmaceae	Bark, root
Curtisia dentata	Cornaceae	Bark, leaf
Eucomis autumnalis	Hyacinthaceae	Bulb
Euphorbia ingens	Euphorbiceae	Latex
Ganoderma lucidum	Bacidiomycetes	Whole plant
Gentiana sp.	Gentianaceae	Root
Gynura pseudochina	Compositae	Root
Hypoxis hemerocallidea	Hypoxidaceae	Corm
Luisia tenuifolia	Orchidaceae	Whole plant
Lyngbya gracilis	Ocillatoriaceae	Fruit
Martynia annua	Martyniaceae	Leaf
Periploca aphylla	Asclepiadaceae	Whole plant- milky juice
Pittosporum viridiflorum	Pittosporaceae	Bark, root
Polygala senega	Polygalaceae	Root
Prunus sp.	Rosaceae	Bark
Psychotria insularum	Rubiaceae	Whole plant
Pterospermum acerifolium	Sterculiaceae	Flower
Rhaphidophora pertusa	Araceae	Stem
Sesamum indicum	Padaliaceae	Seed
Sonchus oleraceus	Compositae	Whole plant
Sutherlandia frutescens	Fabaceae	Stem, leaf, flower, seed
Tetrastigma serrulatum	Vitaceae	Aerial parts
Trapa natans	Trapaceae	Stem
Tricosanthes kirilowi	Cucurbitaceae	Root

and identification can be made and have to be further assessed for their synergistic effects (Pandey and Madhuri, 2009a; Pandey *et al.*, 2013a).

The anticancer and immunostimulatory activities of *Andrographis paniculata* have been reported by many workers. The chemopreventive potential of 80 per cent hydroalcoholic extract (50 and 180 mg/kg/day for 14 days) of *A. paniculata* has been reported (Singh *et al.*, 2001) against chemotoxicity, including carcinogenicity. In this study, the modulatory influence of *A. paniculata* was observed on hepatic and extrahepatic carcinogen metabolizing enzymes (*viz.*, cytochrome P450), antioxidant enzymes, GST content, lactate dehydrogenase (LDH) and lipid peroxidation in Swiss albino mice.

Azadirachta indica (Neem) has been used in buccal carcinogenesis, skin carcinogenesis, prostate cancer, mammary carcinogenesis, gastric carcinogenesis, Ehrlich carcinoma and B16 melanoma. Dietary neem flowers caused a marked increase in GST activity in liver, resulting in a significant reduction in the activities of some hepatic P450-dependent monooxygenases. These results indicate that neem flowers may have chemopreventive potential. In experimental rats, it was interestingly noted that neem flowers resulted in a marked reduction in the tumours of mammary gland (about 35.2 per cent) and liver (61.7 per cent and 80.1 per cent for benign and malignant tumors, respectively), suggesting that the neem flowers contain some chemopreventive agents capable of inhibiting mammary gland and liver carcinogenesis (Tepsuwan *et al.*, 2002). Administration of ethanolic neem leaf extract inhibited the DMBA induced buccal pouch carcinogenesis in hamster, as revealed by the absence of neoplasms. The chemopreventive effects of ethanolic neem leaf extract might be mediated by induction of apoptosis (Subapriya *et al.*, 2005). The modulatory effects of neem leaf with garlic on hepatic and blood oxidant-antioxidant status may play a key role in preventing cancer development at extrahepatic sites (Arivazhagan *et al.*, 2004). The ethanolic extract of neem caused cell death of prostate cancer cells by inducing apoptosis, as evidenced by a dose-dependent increase in DNA fragmentation and a decrease in cell viability (Kumar *et al.*, 2006a).

Camellia sinensis (tea) is one of the most popular beverages in the world. The consumption of tea has been associated with a decreased risk of developing cancers of the ovary, oral cavity, colon, stomach and prostate. This beneficial effect has been attributed to the catechin (a flavonoid) in tea. The biological effects of catechins are due to their strong antioxidant and antiangiogenic activity as well as their potential to inhibit the cell proliferation and modulate carcinogen metabolism (Pandey and Madhuri, 2009a).

The fruits of *Citrus limon* (Nibu) contain flavonoid, flavone, limonoid, limonene, nobiletin and tangeretin. Flavonoid, tangeretin and nobiletin are potent inhibitors of tumour cell growth and can activate the detoxifying P450 enzyme system. Limonoids inhibit tumour formation by stimulating GST enzyme. Limonene (a terpenoid) also possesses anticancer activity. Nibu is used for inhibition of human breast cancer cell proliferation and delaying of mammary tumorigenesis. It is also used in metastasis and leukemia (Heber, 2004).

The derivatives (*viz.*, chlorogenic, dicaffeoylquinic and tricaffeoylquinic acids) of caffeoylquinic acid contained in *Ipomoea batatas* tubers (Shakarkand) have potential cancer chemoprotective effects (Konczak *et al.*, 2004; Matsui *et al.*, 2004). The 4-ipomeanol (a furanoterpenoid) isolated from *I. batatas* has been found to exhibit anticancer activity against non-small cell lung cancer lines (George and Eapen, 2002).

The leaves of *Martynia annua* and barks of *Prunus* sp. (Lakshmi Prabha *et al.*, 2002), and stems of *Rhaphidophora pertusa* (Prajapati *et al.*, 2003) have been used against neck, lung and abdominal cancers, respectively.

Some Important Medicinal Plants for Curing Cancers

A large number of medicinal plants act as anticancer herbs in experimental and/or clinical cancers/tumours of various organs. Some of those cancers are sarcoma, leukaemia, lymphoma and carcinoma. Many reports describe that the anticancer activity of the medicinal plants is because of the presence of certain phytoconstituents, which possess strong antioxidant activities. Thus, consuming a diet rich in antioxidant plant foods will provide a milieu of phytoconstituents, non-nutritive substances in plants that possess health-protective effects (Pandey, 2011b).

Some Important Plants with Anticancer Activity

In this part, 35 anticancer medicinal plants possessing different anticancer activities have been described, as explained by Pandey (2011b) and Pandey *et al.* (2013a).

Aegle marmelos **Correa ex Roxb. (Bel; Family: Rutaceae)-** Lupeol, isolated from pulp and seeds of *A. marmelos*, possesses strong anticancer activity against breast cancer, malignant lymphoma, malignant melanoma, malignant ascites and leukemia. It shows possesses significant antioxidant activity and reduces the side effects of chemotherapy and radiotherapy.

Allium cepa **Linn. (Piyaz/Onion; Family: Liliaceae/Alliaceae)-** Diallyl disulphide, quercetin flavonoid, allicin, allin, and vitamins C and E isolated from the bulb of *A. cepa* detoxify carcinogen, inhibit *Helicobacter pylori* and arrest the cell cycle from S to G_2M phase. Diallyl disulphide inhibits stomach cancer; while quercetin may cure lung and other cancers.

Allium sativum **Linn. (Lasun/garlic; Family: Liliaceae/Alliaceae)-** Sulphur compounds (diallyl sulphide, diallyl disulphide, allyl propyl disulphide) and allicin have been isolated from *A. sativum* bulb. Allicin inhibits the growth of stomach, liver, colon, breast and endometrium cancers; while sulphur compounds inhibit the cancer cells.

Aloe vera **Tourn. ex Linn./*A. barbadensis* Mill. (Ghee-Kunwar/Indian Aloe; Family: Liliaceae)-** Acemannan (a polysaccharide) isolated from root, pulp, leaves or aerial parts of *A. vera* stimulates immune system and shows significant anticancer activity. Emodin and lectins isolated from this herb exhibit strong anticancer and immunoenhancing activities. Aloe-emodin inhibits the growth and spread of stomach cancer, and various sarcomas by inducing apoptosis. Aloe-emodin has selective anticancer activity against neuroectodermal tumours. Alexin B present in it potentially acts against leukemia. Its polysaccharides have strong immunoenhancing and anticancer properties. *A. vera* contains *'super carbohydrates'* that protect against many cancers, particularly hepatic cancer. This herb prevents genesis, regresses growth and prevents metastasis of cancer. *A. vera* stimulates immune system response of the body by activating macrophages and releasing cytokines, *e.g.*, interferon, interleukin and tumour necrosis factor. It has an extraordinary antioxidant profile and reduces the side effects of chemotherapy and radiotherapy. The leaves of *A. vera* contain glycosides-anthracene derivatives or hydroxyanthraquinone derivatives.

Alpinia galanga **Willd. (Barakulanjan; Family: Zingiberaceae)-** Acetoxy-chavicol-acetate isolated from *A. galanga* possesses significant anticancer activity against cancers of breast, lung, stomach, colon and prostate, multiple myeloma, and leukemia. Pinocembrin isolated from this herb inhibits the growth and spread of colon cancer by arresting cell proliferation and inducing apoptosis. Galangin, a flavonoid isolated from *A. galanga*, possesses strong anticancer, antioxidant, antimutagenic and antiinflammatory properties. Galangin protects against breast and prostate cancers.

Andrographis paniculata **Wall. ex Nees (Kiryat/Kalmegh/Creat; Family: Acanthaceae)-** Andrographolide (active diterpine component) isolated from whole plant of *A. paniculata* has immunoenhancing and strong anticancer activity against cancers of breast, ovary, stomach, colon, prostate, kidney and nasopharynx, malignant melanoma, and leukemia. Andrographolide is a potential enhancer of immune system functions like production of WBCs (the defense cells of our body), release of interferon (an antiviral factor) and activity of lymphatic system (the seat of defense system). Andrographolide exerts direct anticancer activity on cancer cells by arresting G_0/G_1 phase of cell-cycle and inducing apoptosis. Dichloromethane fraction of methanolic extract of *A. paniculata* has strong anticancer activity against colon cancer. *A. paniculata* extract is cytotoxic (cell-killing) against cancer cell as seen in human epidermoid carcinoma of skin, lining of nasopharynx and lymphocytic leukemia cells. The chemoprotective potential of *A. paniculata* against chemotoxicity, including carcinogenicity was seen in mice. Thus, *A. paniculata* elicits anticancer, immunostimulant, antioxidant, anti-human immunodeficiency virus infection/ acquired immunodeficiency syndrome (HIV/AIDS), antiinflammatory and antihepatotoxic properties. It enhances the activity of protective liver enzymes and reduces the side effects of chemotherapy and radiotherapy. *A. paniculata* also contains flavonoid and andrographin.

Aphanamixis polystachya **(Wall.) Parker/*Amoora rohituka* Wight and Arn. (Harinhara/Amoora; Family: Meliaceae)-** Amooranin (a triterpene acid), isolated from *A. polystachya* stem bark, inhibits the growth and spread of breast and cervical cancers by arresting G_2M phase of the cell-cycle and by inducing apoptosis. Amooranin and its derivatives are effective in both chemotherapy-sensitive and chemotherapy-resistant cancers. Amooranin has the ability to overcome (reverse) multidrug-resistance in breast cancer, colon cancer and leukemia.

Azadirachta indica **A. Juss./*Melia azadirachta* Linn. (Neem; Family: Meliaceae)-** Stem bark, leaf and flower of *A. indica* contains about 40 different active principles, known as liminoids, that which exhibit immunoenhancing, antioxidant, antimutagenic, anticancer, antimetastatic, antiinflammatory, hepatoprotective, antiulcer, antifungal and antiviral activities. Liminoids regress the growth and spread of various cancers, *e.g.*, breast, lung, liver, stomach, prostate and skin cancers. Nimbolide, a natural triterpenoid, isolated from *A. indica* leaves and flowers inhibits the growth and spread of various cancers, including colon cancer, malignant lymphoma, malignant melanoma and leukemia by inducing apoptosis (programmed cell death, a process that directs the body's immune cells to identify and destroy cancer cells). Nimbolide also prevents metastasis of cancer. Ethanolic extract of *A. indica* inhibits the growth and spread of prostate cancer by inducing apoptosis and its antiandrogenic effect. This herb reduces the side effects of chemotherapy and radiotherapy. *A. indica* also contains polyphenolic myoinositol, dexamethasone, tannin, β sitosterol, nimbin, quercetin and carotene.

Bauhinia variegata **Linn. (Kachnar; Family: Caesalpiniaceae)-** Cyanidin glucoside, malvidin glucoside, peonidin glucoside and kaempferol galactoside, isolated from the root, stem bark and flower of *B. variegata,* inhibit the growth and spread of various cancers, *e.g.*, cancers of breast, lung, liver, oral cavity and larynx, and malignant ascites. *B. variegata* also have significant hepatoprotective activity.

Berberis vulgaris **Linn. (Kashmal; Family: Berberidaceae)-** Its root bark contains berberine, berbamine, chelidonic acid, citric acid, columbamine, hydrastine, isotetrandrine, jacaranone, magnoflorine, oxycanthine and palmatine. Berberine (an isoquinoline alkaloid) has anticancer, immunoenhancing, antioxidant and antiinflammatory properties. Berberine arrests the cancer cell cycle in G_1-phase and induces apoptosis, and hence it possesses strong anticancer activity against prostate cancer, liver cancer and leukemia. This active principle interferes with P-glycoprotein in chemotherapy-resistant cancers. It also increases the penetration of some chemotherapy drugs through the blood-brain-barrier (BBB), thereby enhancing their effect on intracranial tumours. *B. vulgaris* root bark contains three phenolic compounds, *viz.*, tyramine, cannabisin-G and lyoniresinol. Cannabisin-G and lyoniresinol exhibit strong antioxidant activity. Cannabisin-G protects against breast cancer. *B. vulgaris* also inhibits the growth of stomach and oral cavity cancers.

Catharanthus roseus **G. Don/*Vinca rosea* Linn./*Lochnera rosea* (Linn.) Reichb. (Sadabahar/Madagascar Periwinkle; Family: Apocynaceae)-** The whole plant of *C. roseus* contains more than 70 alkaloids, called *'vinca alkaloids'*, *e.g.*, vinblastine, vincristine and their derivatives. The vinca alkaloids arrest cancer cell proliferation by binding to tubulin in the mitotic spindle, *i.e.*, they inhibit microtubule formation and arrests mitosis in metaphase. They also induce apoptosis (programmed cell death) and inhibit angiogenesis (formation of new blood vessels). These alkaloids inhibit the growth and spread of various cancers, including breast, ovary, cervix, lung, colon, rectum, kidney and testis cancers, neuroblastoma, Hodgkin's disease, malignant lymphoma, multiple myeloma, various sarcomas, rhabdomyosarcoma, and leukemia.

Curcuma longa **Linn./*C. domestica* Valeton (Haldi/Turmeric; Family: Zingiberaceae)-** Curcumin (diferuloyl methane) and curcuminoids, isolated from *C. longa* rhizome (tuber) suppress cancer at every step, *i.e.*, initiation, growth and metastasis. Curcumin (pigment colour of haldi) arrests the cancer cells proliferation in G_2/S phase and induces apoptosis (programmed cell death). The curcumin has shown antiinflammatory, antitumour and antioxidant properties. It inhibits angiogenesis, a crucial step in the growth and metastasis of cancer. Curcumin and genistein (isolated from *Glycine max*) act synergistically to inhibit the growth and spread of oestrogen-positive breast cancer. Curcumin acts even in multidrug-resistant breast cancers. It suppresses adhesion of cancer cells, thus preventing metastasis. It inhibits the growth and spread of various cancers, including that of breast, lung, oesophagus, liver, colon, prostate, head, neck and skin. Curcumin is particularly effective in radiotherapy-resistant prostate cancer. It is effective even in advanced stages of cancer. Curcumin showed chemopreventive effect against DEN/ phenobarbital induced-hepatocarcinogenesis in Wistar strain male albino rats. It also protects from stomach and colon cancers. *C. longa* rhizome is also antimutagenic, antioxidant, immunostimulant, antiinflammatory, radioprotective, stimulant, carminative, alterative, blood purifier, hepatoprotective, antiperiodic and tonic. The rhizomes of *C. longa* are also effective in colon, bladder and prostate cancers, intravesical tumour, fibrosarcoma, HCC, oesophagal carcinogenesis, leukemia, stomach papilloma and solid tumours.

Emblica officinalis **Gaertn./***Phyllanthus emblica* **Linn. (Amla/Amlika/Indian Gooseberry; Family: Euphorbiaceae)-** Amla contains ellagic acid, gallic acid, quercetin, kaempferol, emblicanin, flavonoids, glycosides and proanthocyanidins. Phyllembin (from fruit pulp identified as ethyl gallate), tannin (from fruit, bark and leaves), fixed oil, essential oil and phosphatides (from seeds) and leucodelphinidin (from bark) of *E. officinalis* have also been isolated. Ellagic acid is a powerful antioxidant and has the ability to inhibit mutations in genes. Ellagic acid also repairs chromosomal abnormalities. Emblicanins A and B (tannins) possess strong antioxidant and anticancer properties. *E. officinalis* inhibits the growth and spread of various cancers, including cancers of breast, uterus, pancreas, stomach and liver, and malignant ascites. It is highly nutritious and an important source of vitamin C (a powerful antioxidant), phyllembic acid, lipid, emblicol, colloidal complexes, micic acid amino acids and minerals. *E. officinalis* protects against many cancers, particularly liver cancer. It reduces the side effects of chemotherapy and radiotherapy. Amla fruit contains 18 compounds that inhibit growth of gastric, uterine and breast cancers. It enhances NK cell activity in various tumours. Its extract reduced the ascites and solid tumours induced by DLA cells in mice. The extract also increased the life span of tumour bearing animals.

Fragaria vesca **Linn. (Strawberry; Family: Rosaceae)-** Flavonoid, tannin, borneol, ellagic acid, vitamin C, bioflavonoid and chalcone of *F. vesca* leaf and fruit possess antioxidant and anticancer activities. *F. vesca* acts against various cancers.

Ginkgo biloba **Linn. (Maidenhair Tree; Family: Ginkgoaceae)-** Ginkgetin and ginkgolides (A and B), isolated from *G. biloba,* inhibits the growth and spread of various aggressive cancers like invasive ER negative breast cancer, glioblastoma, HCC, and cancers of ovary, colon, prostate and liver by inducing apoptosis. The *G. biloba* extract acts as antioxidant. It reduces the side effects of chemotherapy and radiotherapy.

Glycine max **Merrill/***G. soja* **Sieb. and Zucc./***G. hispida* **Maxim/***Soja max* **Piper (Bhat/Soybean/Soyabean/Soya; Family: Papillionaceae/Fabaceae)-** Isoflavones (*e.g.,* genistein and daidzein) and saponins, isolated from *G. max* beans (seeds), inhibit the growth and spread of various cancers like cancers of breast, uterus, cervix, ovary, lung, stomach, colon, pancreas, liver, kidney, urinary bladder, prostate, testis, oral cavity, larynx, and thyroid. *G. max* is also effective in nasopharyngeal carcinoma, skin cancer, malignant lymphoma, rhabdomyosarcoma, neuroblastoma, malignant brain tumours and leukemia. Isoflavones and saponins possess wide ranging anticancer properties like inhibition of cancer cell proliferation, promotion of cell differentiation and induction of apoptosis. Genistein acts by blocking angiogenesis, acts as a tyrosine kinase inhibitor (the mechanism of action of many new cancer drugs) and induces apoptosis. Genistein is an excellent intracellular antioxidant, which blocks the supply of oxygen and nutrients to cancer cells, thus killing them by starving. Genistein and quercetin have synergistic anticancer effect against ovarian carcinoma. Saponins decrease invasiveness of glioblastoma cells. Anthocyanins of *G. max* induce apoptosis in leukemic cells. *G. max* protects against many cancers, including that of colon, lung and ovary.

Glycyrrhiza glabra **Linn. (Mulathi/Mulhatti/Licorice/Liquorice; Family: Papilionaceae/Fabaceae/Leguminosae)-** Flavonoids (*e.g.*, flavones, flavonols, isoflavones, chalcones, licochalcones and bihydrochalcones) derived from the root, rhizome or whole plant of *G. glabra* possess strong anticancer, antioxidant, antimutagenic, antiulcer, anti-HIV and hepatoprotective properties. Licochalcone-A inhibits the growth and spread of various cancers, particularly androgen-refractory prostate cancer by inducing apoptosis and arresting cancer cells division. Licoagrochalcone possesses strong anticancer activity against cancers of breast, lung, stomach, colon, liver and kidney, and leukemia. Triterpenoid saponins (*e.g.*, glycyrrhizin and glabranin) isolated from *G. glabra* inhibits the growth and spread of lung cancer and fibrosarcomas. Glycyrrhizic acid isolated from this herb protects against aflatoxins (powerful fungal carcinogens of liver). *G. glabra* also contains coumarin, triterpene sterol (β-amerin stigmasterol), eugenol, indole, glycyrrhetinic acid, chalcone glycosides (*viz.*, isoliquiritin and neoisoliquiritin), and liquiritoside (a flavonoside). This herb stimulates the immune system response of body, and protects against colon cancer and oestrogen-positive breast cancer. The rhizomes and roots are also tonic, expectorant, demulcent, laxative and emollient, and used in genito-urinary diseases, coughs, sore throat, catarrhal affections and in scorpion-sting. Licorice is an extract prepared from the dried roots and stems of *G. glabra*. For more than three thousand years, licorice has been used to treat the cancer, hepatitis and some other diseases. Antitumour and antimetastatic effects of cyclophosphamide are potentiated by licorice extract.

Malus domestica **Borkh./***M. pumila* **Mill./***M. communis* **DC/***M. sylvestris* **Hort., non Mill./***Pyrus malus* **Linn. in part (Seb/Sev/Apple; Family: Rosaceae)-** Its fruit (apple) possesses antioxidant and anticancer activities, and may be useful in various cancers.

Morinda citrifolia **Linn. (Al/Ach/Bartundi/Noni; Family: Rubiaceae)-** This has 23 different phytochemicals, including five vitamins and three minerals. The heartwood of *M. citrifolia* contains active constituents as anthraquinones, *viz.*, damnacanthal, rubiadin-methyl ether, alizarin, morindone and anthragallol-2,3-dimethyl ether. Damnacanthal, NB10 and NB11, isolated from the *M. citrifolia* fruit, possess strong anticancer activity against various cancers, particularly lung cancer and sarcomas. *M. citrifolia* possesses strong antioxidant, hepatoprotective and immunoenhancing properties. The flowers on ethanolic extraction yielded acacetin 7-O-(-D(+)-glucopyranoside; 5,7-dimethyl-apigenin-4'-O-(-D(+) galactopyranoside and a new anthraquinone glycoside. The fruit juice of *M. citrifolia* showed antitumour activity against intraperitoneally implanted Lewis lung carcinoma in syngenic mice. Noni fruit extract acts indirectly on cancer cells by enhancing the host immune system. There is a polysaccharide compound (6-D-glucopyranone pentaacetate) found in Noni that increases the ability of immune system to produce chemicals which enhance the killing power of WBCs against cancer. Noni fruit provides a safe and effective way to increase xeronine levels, which exert a crucial influence on cell health and body protection. Its fruit contains proxeronine (a precursor of xeronine), which initiates the release of xeronine in the intestinal tract after it comes in contact with a specific enzyme (present in fruit). Thus, xeronine is an alkaloid to which the body produces in order to activate enzymes, so they can function properly. This particular alkaloid has never been found because the body makes it, immediately uses it, and then breaks it down. Xeronine is so basis to the functioning of proteins, we would die without it. Its absence can cause many kinds of illness. Noni, which is probably the best source of proxeronine, acts as an immunostimulant, inhibits the growth of certain tumours, enhances and normalizes cellular functions, and boosts tissue regeneration.

Nigella sativa **Linn. (Kalonji/Kalajira/Black Cumin; Family: Ranunculaceae)-** Thymoquinone and dithymoquinone, isolated from *N. sativa* seeds, have strong anticancer activity against various cancers, including cancers of colon, prostate, pancreas and uterus, malignant ascites, malignant lymphoma, malignant melanoma, sarcomas, and leukemia. Thymoquinone is effective in both hormone-sensitive and hormone-refractory prostate cancers. *N. sativa* kills cancer cells by binding to the asialofeutin (lectin) on the surface of cancerous cells, causing their aggregation and clumping. It also possesses immunoenhancing and antiinflammatory properties. It protects against liver cancer. *N. sativa* enhances the immune function of the body and reduces the side effects of chemotherapy and radiotherapy.

Ocimum sanctum **Linn. (Tulsi/Sacred Basil/Holy Basil; Family: Labiatae/ Lamiaceae)-** Its leaves contain volatile oils (comprising of eugenol and methyl eugenol), linolenic acid, oleanolic acid, rosmarinic acid, and flavonoids or phenolic compounds as antioxidants (*e.g.*, orientin, vicenin, cirsilineol, cirsimaritin, isothymusin, isothymonin and apigenin). The volatile oils also contain carvacrol and sesquiterpene hydrocarbon caryophyllene. Ursolic acid, apigenin, luteolin, apigenin-7-O-glucuronide, luteolin-7-O-glucuronide, orientin and molludistin have also been isolated from the leaves. *O. sanctum* also contains a number of sesquiterpenes and monoterpenes, *viz.*, bornyl acetate, β-elemene, neral, α- and β-pinenes, camphene, campesterol, cholesterol, stigmasterol, and β sitosterol. Eugenol, orientin and vicenin inhibit the growth and spread of various cancers such as breast cancer, liver cancer and sarcomas, particularly fibrosarcoma by blocking supply of oxygen and nutrients to cancer cells and killing them by starving.

Ursolic acid has immunoenhancing and tissue-protective properties. Polysaccharides isolated from *O. sanctum* have antioxidant and radioprotective properties. It possesses antioxidant, antitumour and immunomodulatory activities and protects against various cancers, particularly breast cancer and reduces the side effects of chemotherapy and radiotherapy. The AlE of *O. sanctum* leaves has a modulatory influence on carcinogen metabolizing enzymes, *e.g.*, cytochrome P450, cytochrome b_5, aryl hydrocarbon hydroxylase and GST, which are important in detoxification of carcinogens and mutagens. *O. sanctum* significantly decreased the incidence of B(a)P induced neoplasia of fore-stomach of mice and 3'-methyl-4-dimethylaminoazo-benzene induced hepatomas in rats. The AlE of *O. sanctum* leaves showed an inhibitory effect on chemically induced skin papillomas in mice. The leaf extract of *O. sanctum* blocks or suppresses the events associated with chemical carcinogenesis by inhibiting metabolic activation of carcinogen.

Oldenlandia diffusa Roxb./*Hedyotis diffusa* Willd. (Family: Rubiaceae)- This is a native of China. The whole plant of *O. diffusa* contains oldenlandosides, stigmasterol, ursolic acid, oleanolic acid, β sitosterol, p-coumaric acid and flavonoid glycosides. Ursolic acid inhibits the growth and spread of various cancers such as cancers of lung, ovary, uterus, stomach, liver, colon, rectum and brain, malignant melanoma, malignant ascites, lymphosarcoma, and leukemia. Ursolic acid works by a typical cytotoxic effect on cancer cells and by inducing apoptosis.

Panax ginseng Mey./*P. schinseng* Nees (Asiatic or Chinese Ginseng; Family: Araliaceae)- Ginsenosides (panaxadiol and panaxatriol saponins), isolated from *P. ginseng*, inhibits the growth and spread of various cancers like cancers of breast, ovary, lung, prostate and colon, renal cell carcinoma, malignant melanoma, malignant lymphoma, and leukemia. Panaxadiol ginsenosides (Rb_1, Rb_2, Rc, Rd, Rg_3, Rh_2) and panaxatriols ginsenosides (Re, Rf, Rg_1, Rg_2, Rhi) have both preventive and therapeutic roles in cancer treatment. Ginsenosides possess strong anticancer activity against lung cancer, and also prevent lung metastasis by blocking angiogenesis. Compound K (a metabolite of ginsenosides) inhibits the growth and spread of chemo-resistant lung cancer. Ginsenosides Rc, Rd, Rg_1 and Re overcome (reverse) P-glycoprotein mediated multi-drug resistance to chemotherapy. Ginsenoside Rf helps in reducing doses of morphine in terminally ill-cancer patients. Polysaccharides of *P. ginseng* possess strong immunoenhancing and anticancer activities against many cancers, particularly lung cancer. These polysaccharides also reduce the side effects of chemotherapy and radiotherapy. *P. ginseng* also possesses antistress, hepatoprotective, haemopoietic, immunoenhancing, antioxidant, radioprotective, chemoprotective and antiinflammatory properties. It inhibits proliferation and seeding (metastases) in various cancers by inducing cell differentiation and apoptosis. It is effective in both hormone-responsive and hormone-refractory prostate and breast cancers.

Plumbago zeylanica Linn. (Chitrak/Chitra; Family: Plumbaginaceae)- Plumbagin, isolated from *P. zeylanica* root inhibits growth and spread of breast cancer, liver cancer, fibrosarcoma, malignant ascites and leukaemia by inhibiting cancer cell proliferation. *P. zeylanica* also possesses strong antioxidant, hepatoprotective, neuroprotective and immunoenhancing properties.

Podophyllum hexandrum Royle/*P. emodi* Wall. ex Hook. f. and Thoms./ *P. peltatum* (Papra/Indian Podophyllum/Himalayan May Apple; Family: Berberidaceae)- Podophyllotoxin and podophyllin (lignans) isolated from *P. hexandrum* inhibit the growth and spread of various cancers, including cancers of breast, ovary, lung, liver, urinary bladder, testis and brain, neuroblastoma, Hodgkin's disease, non-Hodgkin's lymphoma, and leukemia. Podophyllotoxin is the most active among all the natural anticancer compounds. *P. hexandrum* also possesses potent radioprotective and haemopoietic properties.

Prunella vulgaris Linn./*Brunella vulgaris* Linn. (Dharu; Family: Labiatae/ Lamiaceae)- Ursolic and oleanolic acids, isolated from *P. vulgaris*, inhibit the growth and spread of various cancers like cancers of breast, cervix, lung, oral cavity, oesophagus, stomach, colon and thyroid, malignant lymphoma, intracranial tumours and leukemia. This herb also possesses immunoenhancing, hepatoprotective, antioxidant, anti-HIV and anti-Herpes properties. It has normoblastic effect on the bone marrow.

Psoralea corylifolia **Linn. (Babchi; Family: Papilionaceae/Fabaceae)-** Bavachinin, corylfolinin and psoralen, isolated from *P. corylifolia*, possess strong anticancer activity against lung cancer, liver cancer, osteosarcoma, fibrosarcoma, malignant ascites and leukemia. Psoralen enhances the body immunity by stimulating NK cell activity. Psoralidin isolated from this herb inhibits the growth and spread of stomach and prostate cancers by inhibiting G_2M phase of cell-cycle. Psoralidin induces apoptosis in both androgen-responsive and androgen-refractory prostate cancers. *P. corylifolia* also possesses strong antioxidant, immunomoenhancing and hepatoprotective properties.

Punica granatum **Linn. (Anar/Pomegranate; Family: Punicaceae)-** Fruit (Anar) of *P. granatum* contains alkaloids, anthocyanidines and vitamin C. It acted against solid tumour and ascites tumour in albino mice.

Rubia cordifolia **Linn. sensu Hook. f. (Manjit/Majith/Rosemary; Family: Rubiaceae)-** Its root contains rubidianin, rubiadin, rosemary acids (*viz.,* RA-7, RA-700, RC-1 and RC-18), carnosic acid, purpurin, pseudopurpurin, alizarin and xanthopurpurin (purpuroxanthin). The *R. cordifolia* root inhibits the growth and spread of breast, ovary, cervix, colon and lung cancers, malignant ascites, malignant lymphoma, malignant melanoma (B16 melanoma), P388 cells, L1210 cells, sarcoma, and leukemia. Rubiadin also possesses hepatoprotective activity.

Saussurea lappa **C.B. Clarke (Kut/Kur/Kuth/Costus; Family: Compositae/ Asteraceae)-** Sesquiterpenes and costunolide dehydrocostuslactone of *S. lappa* inhibit the growth and spread of breast cancer. Cynaropicrin from *S. lappa* possesses strong anticancer activity against malignant lymphoma and leukemia. Costunolide also inhibits the growth and spread of intestinal cancer. Mokkolactone content of this herb induces apoptosis in leukemic cells. Shikokiols of *S. lappa* exhibit anticancer activity against cancers of ovary, lung, colon and CNS. *S. lappa* inhibits growth and spread of cancers by arresting cancer cell division in G_2 phase, and by inducing apoptosis.

Solanum nigrum **Linn. (Makoi/Kakmachi/Vayasi/Black nightshade; Family: Solanaceae)-** Flavonoids (*e.g.,* quercetin) and alkaloids (*viz.,* solasodine, solanine and solamargine) are the main phytoconstituents of *S. nigrum* whole palnt or fruit. These constituents have been reported to act against various tumours. Solamargine and solasonine inhibit the growth and spread of various cancers, including breast, liver, lung and cyst cancers, choriocarcinoma or chorioadenoma, and leukemia. Solanine and solamargine have very strong anticancer actions against murine tumours. Steroidal glycosides (spirostane, furostane, spirosolane and pregnane), isolated from *S. nigrum*, inhibit the growth and spread of colon cancer and pheochromocytoma. Its glycoproteins have antiproliferative and apoptotic effects on colon and breast cancers. Its polysaccharides have significant inhibitory effect on the growth of cervical cancer. *S. nigrum* inhibits the growth and spread of liver cancer by two distinct anticancer activities, *i.e.*, apoptosis (programmed cell death) and autophagy (autophagocytosis). Higher doses of *S. nigrum* induce apoptotic cell death, while lower doses lead to autophagocytic death of cancer cells. Lunasin of *S. nigrum* is a cancer-preventive peptide. *S. nigrum* and *S. lyrati* inhibit the growth and spread of stomach cancer, sarcomas, malignant ascites and leukemia. Its leaf extract has inhibitory effect against S180, V14 and Ec tumour models.

Tinospora cordifolia **(Willd.) Miers ex Hook. f. and Thoms. (Giloe/Amrita/ Gulancha/Gulbel/Tinospora; Family: Menispermaceae)-** Its stem bark and fruit contain berberine, tinosporine, giloin and giloinin. Sesquiterpenes and diterpenes of this herb inhibit the growth and spread of various cancers, including cancers of lung, cervix and throat, and malignant ascites. Polysaccharide fraction of *T. cordifolia* inhibits lung metastasis. Arabinogalactan, syringine, cordiol, cordioside, cordifoliosides (A and B) obtained from *T. cordifolia* possesses significant immunoenhancing activity. *T. cordifolia* reduces the side effects of radiotherapy and chemotherapy. This herb also possesses antioxidant, neuroprotective, hepatoprotective, antistress, antiulcer, antiasthmatic, antidiabetic (or hypoglycaemic and hypolipidaemic) and antipyretic activities.

Viscum album **Linn./*V. costatum* Gamble (Banda/Ban/European Mistletoe; Family: Loranthaceae/Viscaceae)-** The whole plant of *V. album* contains lectin alkaloids, acetylcholine, proprionyl choline, lupeol, viscotoxin, flavonoid and sterol A. Lectins (*e.g.*, viscumin), polypeptides (viscotoxins) and phenolic compounds (*e.g.*, digallic acid), isolated from *V. album* inhibit the growth and spread of various cancers, including that of breast, cervix, ovary, lung, stomach, colon, rectum, kidney, urinary bladder and testis, malignant melanoma, sarcoma, fibrosarcoma, malignant ascites, lung metastasis, and leukemia by inducing apoptosis and antiangiogenesis activity. Lectins possess both anticancer and immunostimulating activities. Lectin-II induces apoptosis in cancer cells via activation of caspase-3 cascades. Lectin alkaloids also cause lectin II induced apoptosis and inhibition of telomerase via mitochondrial controlled pathway independent of p53. Hexamethylene bioacetamide of *V. album* causes p53-dependent apoptosis and induction with telomerase. Viscumin, responsible for most of the biological activities of *V. album*, acts by bringing together immune system effector cells and cancer cells.

Withania somnifera **Dunal (Ashwagandha/Asgandh/Punir; Family: Solanaceae)-** Majority of the phytoconstituents of *W. somnifera* root are withanolides (steroidal lactones with ergostane skeleton) and alkaloids. They include withanone, withaferin A, and several other withanolides and withasonidienone. Apart from these, *W. somnifera* root also contains withaniol, acylsteryl glucosides, starch, glycosides, reducing sugar, resins, saponins, fixed oils, hentriacotane, ducitol, anthraquinones, proteins, amino acids (*e.g.*, aspartic acid, proline, tyrosine, alanine, glycine, glutamic acid, cystine and tryptophan) and high amount of iron, etc. Withaferin A and withanolide D have antioxidant, anticancer and immunoenhancing activities, and act against various cancers. Withanolides are similar to ginsenosides (the active principles of *P. ginseng*) in both structure and activity. Withanolides (including withaferin A, sitoindoside IX, physagulin D, withanoside IV and viscosalactone B) inhibit the growth and spread of various cancers such as cancers of breast, lung, colon and CNS due to their antiproliferative and antiangiogenic properties. Withaferin A (the most important withanolide) inhibits the growth and spread of various cancers, including that of breast, cervix, colon, prostate, nasopharynx and larynx, malignant ascites, and sarcoma by inducing apoptosis. Withaferin A is effective in both androgen-responsive and androgen-refractory prostate cancers. Sitoindosides VII-X and withaferin A have strong antioxidant, antistress, immunomodulatory, antiinflammatory and antiaging properties. Withanolide D inhibits the metastatic colony formation in malignant lung melanoma. Ashwagandhanolide, a new dimeric withanolide, isolated from *W. somnifera*, inhibits the growth and spread of breast, stomach, colon, lung and CNS cancers. *W. somnifera* reduced the cancer cell proliferation and increased the overall survival time. It enhanced the effectiveness of radiation therapy, and reduced the side effects of radiotherapy and chemotherapy. Given its broad spectrum of cytotoxic and anticancer activity, *W. somnifera* is a novel therapy for cancer.

Zingiber officinale **Rosc. (Adrak/Ada/Ginger; Family:** *Zingiberaceae*)- Gingerols, isolated from *Z. officinale* rhizome inhibit the growth and spread of various cancers, including that of ovary, cervix, colon, rectum, liver, urinary bladder and oral cavity, neuroblastoma, and leukemia by inducing apoptosis. The most active individual component, 6-shogaol, isolated from *Z. officinale*, inhibits the growth and spread of many cancers, particularly the ovarian cancer by blocking formation of new blood vessels, and by inducing apoptosis and autophagy. It is effective even in chemotherapy-resistant ovarian cancer. *Z. officinale* reduces the side effects of chemotherapy and radiotherapy. It also possesses antioxidant, antimutagenic and antiinflammatory activities.

Some Anticancer Plants of Foreign Origin

Cancer is the abnormal growth of cells in our bodies that can lead to death. Cancer cells usually invade and destroy normal cells. These cells born of an imbalance in the body and by correcting imbalance, the cancer may go away. Billions of dollars have been spent on cancer research and yet we still don't understand exactly what cancer is. Every year, millions of people are diagnosed with cancer, leading to death. Cancer is the second leading cause of death in America. The major causes of cancer are smoking, dietary imbalances, hormones and chronic infections, leading to chronic inflammation (Madhuri and Pandey, 2009a). According to the American Cancer Society (2006), deaths arising from cancer constitute 2 to 3 per cent of the annual deaths recorded worldwide. Breast cancer is the most common form of cancer in women worldwide. Amongst the South African women, breast cancer is likely to develop in one of every woman in this country (Koduru *et al.*, 2007). Colon cancer is the second most common cause of cancer deaths in the United State. Prostate cancer is the most frequently diagnosed cancer among men in the US, second to skin cancer with an estimated 180,000 new cases and 37,000 deaths expected by American Cancer Society each year. With increase in longevity, it is going to be a problem even in India. Cancers affecting the digestive tract are among the most common of all the cancers associated with ageing. About one of every 14 men and women in America are diagnosed with a gastro-intestinal cancer at some time in their lives (Madhuri and Pandey, 2009a; Pandey *et al.*, 2013a).

Because of high death rate associated with cancer and because of the serious side effects of chemotherapy and radiation therapy, many cancer patients seek out alternative and or complementary methods of treatment. The important preventive methods for most of the cancers include dietary changes, stopping the use of tobacco products, treating inflammatory diseases effectively, and taking nutritional supplements that aid immune functions. Over the past decade, herbal medicines have been accepted universally, and they put the impact on both world health and international trade. Hence, the medicinal plants continue to play important role in the healthcare system of large number of world's population (Madhuri and Pandey, 2009a). Traditional medicine is widely used in India. Even in USA, use of plants and phytomedicines has increased dramatically in last two decades. A National Centre for Complementary and Alternative Medicine has been established in USA. The herbal products have been classified under *'Dietary Supplements'* and are included with vitamins, minerals, amino acids and 'other products intended to supplement the diet' (Rao *et al.*, 2004). Use of plants for medicinal remedy is an integral part of the South African cultural life. It is estimated that 27 million South Africans use herbal medicines from more than 1020 plant species (Pandey *et al.*, 2013a).

Certain Anticancer Plants of Foreign Native

The data on 62 medicinal plants of foreign origin are presented in Table 10. Many of these medicinal plants have been found very effective in experimental as well as clinical cases of cancers (Madhuri and Pandey, 2009a; Pandey *et al.*, 2013a).

Many naturally occurring substances present in the human diet have been identified as potential chemopreventive agents; and consuming relatively large amounts of vegetables and fruits can prevent the development of cancer (American Cancer Society, 2006; Vecchia and Tavani, 1998). Compared with meat eaters, most, but not all, studies have found that the vegetarians are less likely to be diagnosed with cancer. Vegetarians have also been shown to have stronger immune function, possibly explaining why vegetarians may be partially protected against cancer (Madhuri and Pandey, 2008). Many plant-derived products have been reported to exhibit potent antitumour activity against several rodent and human cancer cell lines (Madhuri and Pandey, 2009a).

The phytochemicals, *e.g.*, vitamins (A, C, E, K), carotenoids, terpenoids, flavonoids, polyphenols, alkaloids, tannins, saponins, pigments, enzymes and minerals, etc. have been found to elicit antioxidant activities (Heber, 2004; Kathiresan *et al.*, 2006; Kaur and Kapoor, 2002). Ellagic acid and a whole range of flavonoids, carotenoids and terpenoids present in *Fragaria vesca* (strawberry) and *Rubus idaeus* (raspberry) act as antioxidants. These chemicals block the action of various hormones and metabolic pathways that are associated with the development of cancer. *Rosmarinus officinalis* (rosemary) contains substantial amounts of carnosol and ursolic acid, the potent antioxidants that possess antitumor activity. Quercetin is the major flavonol in the western diet. Rich sources of quercetin are red and yellow onions, kale, broccoli, red grapes, cherry, French bean, apple and cereals. Quercetin possesses both anticarcinogenic activity and the ability to inhibit the oxidation of low-density lipoprotein (LDL) cholesterol oxidation. A whole variety of phenolic compounds, in

Table 10: Certain Anticancer Plants of Foreign Origin

Botanical Name of Plant (with Family)	Parts Used and their Main Active Components	Origin/Native Place
Agave americana (Agavaceae)	Leaf contains steroidal saponin, alkaloid, coumarin, isoflavonoid, hecogenin and vitamins (A, B, C)	Central America
Agrimonia pilosa (Rosaceae)	Herb contains agrimonolide, flavonoid, triterpene, tannin and coumarin	China, Japan, Korea, India
Agropyron repens (Poaceae)	Rhizome contains essential oil, polysaccharide and mucilage	Europe
Ailanthus altissima (Simaroubaceae)	Bark contains triterpene, tannin, saponin and quercetin 3-glucoside	China, Korea
Akebia quinata (Lardizabalaceae)	Fruit contains flavonoid and saponin	China, Japan, Korea
Alpinia galangal (Zinziberaceae)	Rhizome contains kaempferide and flavone	Europe
Aristolochia contorta (Aristolochiaceae)	Root and fruit contain lysicamine and oxaaporphine	China, Korea
Aster tataricus (Asteraceae)	Whole plant and root contain triterpene, monoterpene and epifriedelanol	Japan, Korea
Broyonia dioica	Root contains cucurbitacin and glycoside.	Europe
Cannabis sativa (Cannabinaceae)	Leaf contains stereo isomers of cannabitriol	South Africa
Chelidonium jajus var. asiaticum (Papaveraceae)	Herb contains alkaloids (sanguinarine, chelerythrine, berberine)	Asia, Europe
Chimaphila umbellate (Ericaceae)	Whole plant contains ericolin, arbutin, urson and tannin	Asia, Europe
Coix lacryma-jobi (Poaceae)	Seed contains trans-ferulyl stigmasterol	China
Dryopteris crassirhizoma (Polypodiaceae)	Rhizome contains filicinic and filicic acids, aspidinol and aspidin	China, Japan, Korea
Echinops setifer (Asteraceae)	Whole plant contains echinopsine	Korea
Erythronium americanum (Liliaceae)	Whole plant contains α-methylene-butyrolactone	North America
Euonymus alatus (Celastraceae)	Whole plant contains triterpene, euolatin, steroid and sesquiterpene alkaloid	China, Japan, Korea
Eupatorium cannabinum (Asteraceae)	Whole plant contains sesquiterpene, lactone, pyrrolizidine alkaloid and flavonoid	Europe, Asia, North America

Contd...

Table 10–Contd...

Botanical Name of Plant (with Family)	Parts Used and their Main Active Components	Origin/Native Place
Fragaria vesca (Rosaceae)	Leaf and fruit contain flavonoid, tannin, borneol and ellagic acid	Asia, Europe
Fritillaria thunbergii (Liliaceae)	Whole plant contains alkaloid and peimine	China, Siberia
Galium aparine (Rubiaceae)	Cleaver contains iridoid, polyphenolic acid, tannin, anthraquinone and flavonoid	Europe, Africa, Australia
Hydrastis canadensis (Ranunculaceae)	Whole plant contains isoquinoline alkaloids (hydrastine, berberine, berberastine, candaline), resin and lactone	Canada, United States
Hypoxis argentea (Hypoxidaceae)	Corm	South Africa
Junchus effuses (Juncaceae)	Whole plant contains tridecanone, effusol, juncanol, phenylpropanoid and α-tocopherol	China, Japan, Korea
Knowltonia capensis (Ranunculaceae)	Leaf	South Africa
Lantana camara (Verbenaceae)	Whole plant contains alkaloids (camerine, isocamerine, micranine, lantanine, lantadene)	Tropical America
Larrea tridentate (Zygophyllaceae)	Whole plant contains resin	South western USA, Mexico
Lonicera japonica (Caprifoliaceae)	Whole plant, stem and flower contain tannins, saponins and carotenoid	China
Merwilla plumbea (Hyacinthaceae)	Bulb	South Africa
Nidus vespae	Whole plant	China
Olea europaea (Oleaceae)	Leaf and oil contain oleic acid and polyphenol	America
Oldenlandia diffusa (Rubiaceae)	Whole plant	China
Panax quinquefolius (Araliaceae)	Root contains ginsenoside, sesquiterpene, limonene and vitamins (B_1, B_2, B_{12})	China, Japan, Korea
Patrinia heterophylla (Vlerianaceae)	Whole plant	China
Patrinia scabiosaefolia (Vlerianaceae)	Whole plant	China, Japan, Korea
Phaleria macrocarpa	Fruit contains gallic acid	Indonasia
Polygonatum multiflorum (Liliaceae)	Whole plant contains saponin, flavonoid and vitamin A	Asia, Europe, North America
Polygonum cuspidatum (Polygonaceae)	Whole plant	China

Contd...

Table 10–Contd...

Botanical Name of Plant (with Family)	Parts Used and their Main Active Components	Origin/Native Place
Potentilla chinensis (Rosaceae)	Whole plant contains gallic acid and tannin	China, Japan, Korea
Pteris multifida	Whole plant	China
Pygeum africanum (Boraginaceae)	Bark contains phytosterol, triterpene and tannin	Africa
Pyrus malus (Rosaceae)	Bark and fruit contain quercetin, catechin, flavonoid, coumaric and gallic acids, phloridzin and procyanidin	Britain
Rhus chinensis (Anacardiaceae)	Leaf contains tannin, apigenin and glycoside; seed contains bruceosides (A, B), brucein D and fatty oil	China, Japan, Korea
Rosmarinus officinalis (Lamiaceae)	Whole plant contains volatile oil, borneol, carnosol, ursolic acid, diterpene, rosmaricine, flavonoid and tannin	South Europe
Rubia akane (Rubiaceae)	Whole plant contains anthraquinone and triterpene	Japan, Korea
Rubus idaeus (Rosaceae)	Leaf contains flavonoid and tannin; fruit contains vitamins (A, B, C) and ellagic acid	Asia, Europe
Scilla natalensis (Hyacinthaceae)	Bulb	South Africa
Scrophularia nodosa (Scrophulariaceae)	Aerial part contains iridoid, flavonoid and phenolic acid	Europe
Scutellaria barbata (Lamiaceae)	Whole plant	China
Smilax chinensis (Liliaceae)	Rhizome contains tannin, saponins and flavonoid	China, Japan
Smilax glabra (Liliaceae)	Rhizome	China
Solanum aculeastrum (Solanaceae)	Root bark, leaf and fruit	South Africa
Solanum lyrati (Solanaceae)	Whole plant	China
Sophora flavescens (Fabaceae)	Root	China
Sophora subprostrata (Fabaceae)	Root	China
Tabebuia sp. (Bignoniaceae)	Bark contains quinine, bioflavonoid and co-enzyme Q	South America
Taraxacum mongolicum (Asteraceae)	Whole plant	China, Korea, Mongolia

Contd...

Table 10—Contd...

Botanical Name of Plant (with Family)	Parts Used and their Main Active Components	Origin/Native Place
Thuja occidentalis (Cupressaceae)	Whole plant contains flavonoid, tannin, volatile oil and mucilage	North eastern USA, Europe
Thymus vulgaris (Lamiaceae)	Whole plant contains volatile oil, flavonoid and tannin	South Europe
Trifolium pratense (Fabaceae)	Flower contains glucosides (trifolin, trifolitin, trifolianol), flavonoid and phenolic acid	Asia, Europe, Africa, Australia
Tulbaghia violacea (Alliaceae)	Bulb	South Africa
Vitex rotundifolia (Verbenaceae)	Whole plant contains camphene, pinene and diterpene	China, Japan, Korea

addition to the flavonoids, are widely distributed in grains, fruits, vegetables and herbs. The phenolic compounds like caffeic, ellagic and ferulic acids, sesamol and vanillin have been reported to exhibit antioxidant and anticarcinogenic activities and inhibit atherosclerosis (Madhuri and Pandey, 2009a).

The medicinal plants cited in Table 10 have been found very effective in various types of malignant (cancer) and benign tumours of humans and experimental animals. For examples: *A. pilosa* in S180; *A. altissima* in intestinal cancer, S180, sarcoma 37 and leukemia 16; *A. quinata* in S180 and sarcoma 37; *C. jajus* var. *asiaticum* in stomach cancer; *C. umbellate* in breast tumour; *C. lachryma-jobi* in ascites cancer and Yoshida's sarcoma; *F. thunbergii* in tumours of throat, chest, neck and breast; *L. tridentate* in various cancers, especially leukemia; *L. japonica* in ascites carcinoma and S180; *N. vespae* in gastric and liver cancers; *O. diffusa* in leukemia, Yoshida's sarcoma, S180 and Ehrlich's ascites sarcoma; *P. heterophylla* and *P. scabiosaefolia* in ascites cancer; *P. macrocarpa* in oesophageal cancer; *P. cuspidatum* in S180; *P. multifida* in S180, sarcoma 37 and Yoshida's sarcoma; *P. africanum* in prostate cancer; *P. malus* in lung, colon, breast and intestinal cancers; *S. barbata* in S180 and EAC; *S. chinensis* and *S. glabra* in S180 and ascites sarcoma; *S. lyrati* in S180, sarcoma 37, EAC and stomach cancer; *S. flavescens* and *S. subprostrata* in S180, leukemia and cervical cancer 14; *T. mongolicum* in ascites cancer, S180 and lung cancer cells; and *V. rotundifolia* in lung tumour.

Chapter 5
Anticancer Plant-Drugs: Part II

The Anticancer Dietary Plants

Use of plants for treating various diseases of both animal and man is old practice. The agricultural plants not only maintain the health and vitality of individuals, but also cure various diseases, including cancer without causing any toxicity. Some agricultural plants and their products such as vegetables, fruits and crops play an important role for the prevention of cancer in our life. Many naturally occurring substances present in the diet/food have been identified as potential chemopreventive agents. A published study of 2400 Greek women noted that consumption of vegetables and fruits were independently associated with significant reductions in the incidence of breast cancer. Vegetarians have been shown to have stronger immune function, possibly explaining why vegetarians may be partially protected against cancer (Madhuri and Pandey, 2008).

Certain Dietary Agricultural Plants as Anticancer

The data on certain some dietary (food) agricultural plants are presented here to explore out their anticancer and immunomodulatory properties. These plants are commonly consumed in our diet as food matters, and are also helpful for the livelihood of animals (Madhuri and Pandey, 2008; Pandey *et al.*, 2013a).

Allium cepa **Linn. (Piyaz)-** The bulb of *A. cepa* contains diallyl disulphide (a sulphurous product), quercetin flavonoid, allicin, allin, amino acids and vitamins (folic acid, C, E). Diallyl disulphide inhibits cancerous cells in stomach. Quercetin, because of its antioxidant properties, may treat lung cancer and other cancers.

Allium cepa **var.** *aggregatum* **(Shallot)-** Its bulb contains allyl propyl disulphide (a sulphurous product), flavonoid, phenolic content, proteins and vitamins (folic acid, C). Allyl propyl disulphide, phenolic content and flavonoid inhibit cancerous cells in liver, stomach and other organs. It helps liver to eliminate toxins to inhibit and kill cancer cells.

Allium sativum **Linn. (Lasun)-** Its bulb contains sulphur compounds (diallyl sulphide, diallyl disulphide, allyl propyl disulphide), allicin, allin and amino acids. The sulphur compounds inhibit cell proliferation, modulate cell cycle activity and interfere with hormone action in cancer cells. Allicin inhibits proliferation of cancerous cells of human mammary gland, endometrium and colon. Lasun is also used in gastric and liver cancers.

Brassica campestris **Linn. (Sarson)-** The seed oil of *Br. campestris* contains glycerides of palmitic, oleic and linoleic acids. It is used in various tumours.

Brassica oleracea **var.** *botrytis* **Linn. (Phoolgobhi)-** Its sprout contains cysteine, ascorbigen, sulphoxide, indole-3-carbinol, glucaric acid, sulphoraphane, glucosinolates, isothiocyanates and vitamins (A, C). These components possess anticancer activities in lung, stomach, colon, bladder, mammary and rectum cancers.

Brassica oleracea **var.** *capitata* **Linn. (Cabbage or Pattagobhi)-** Cabbage contains glucaric acid, sulphoraphane, glucosinolates, isothiocyanates, cysteine, ascorbigen, sulphoxide, indole-3 carbinol, allyl isothiocyanate and vitamins (A, B, C). These have anticancer activities against bladder, lung, stomach, colon, rectum and mammary cancers.

Brassica rapa **Linn. (Shalgam)-** Its leaf and root contain ascorbigen and vitamins (A, C), and are used in various tumours.

Citrus limon **Linn. (Baranibu)-** The fruit (lemon) contains flavonoid, flavone, limonoid, limonene, nobiletin and tangeretin. Flavonoid, tangeretin and nobiletin are potent inhibitors of tumour cell growth, and can activate the detoxifying P-450 enzyme system. Limonoids inhibit tumour formation by stimulating the enzyme GST. Limonene (a terpenoid) also possesses anticancer activity. Nibu fruit is used for inhibition of human breast cancer cell proliferation and delaying of mammary tumorigenesis. It is also used in metastasis and leukemia.

Curcuma longa **Linn. (Haldi)-** Its rhizome (Haldi) contains curcumin, zingiberine and curcuminoids. It is used in colon, bladder and prostate cancers, intravesical tumour, fibrosarcoma, HCC, oesophagal carcinogenesis, leukemia, stomach papilloma and solid tumour cell lines.

Emblica officinalis **Gaertn. (Amla)-** Its fruit (Amla) contains ascorbic acid (vitamin C) and phyllembic acid. It inhibits cellular mutation and may prevent cancer. Its 18 compounds inhibit the growth of gastric and uterine cancer cells. Amla also inhibited the growth of *in vitro* human breast cancer cells.

Glycine javanica **Linn. (Soybean)-** The soybean seed contains phytates, protease inhibitors, phytosterols, saponins, isoflavonoids and isoflavones. Soybean is the contributing factor in the low incidence of breast, prostate, stomach, colon, rectum and lung cancers. Its isoflavonoids inhibit the growth of hormone-dependent and hormone-independent cancer cells in culture. Isoflavones inhibit the growth of human breast and prostate cancers.

Lycopersicon esculentum **Mill. (Tamatar)-** Its fruit (Tamatar) contains lycopene, which is an antioxidant. Lycopene inhibits the proliferation of cancer cells. Higher intake of tamatar or higher blood levels of lycopene is correlated with protection from the cancer; the protective effect was strongest for cancers of prostate, lung and stomach but some protective effects also appeared for cancers of pancreas, colon, rectum, oesophagus (throat), mouth, breast and cervix.

Momordica charantia **Linn. (Karela)-** Its leaf, fruit and seed contain linolenic acid, momordin, palmitic acid, proteolytic enzymes and vitamins (A, C). These are used in colon, mammary and bladder carcinomas, lymphoma, leukemia and other tumours.

Swertia chirata **Buch.-Ham. (Chirayita)-** The whole plant of *S. chirata* contains swertianin, chiratanin, swertinin and swertenol. It is used in various tumours.

Trigonella foenumgraecum **Linn. (Methi)-** Its leaf and seed contain choline, trigonelline, amino acids, proteins, vitamins and quercetin. These are used in various cancers.

Triticum aestivum **Linn. (Gehun)-** The grains of *Tr. aestivum* contains ellagic, linolenic and oleic acids, sterols, phytases, tocotrienols and vitamin E. Whole grains protect against lymphomas and cancers of pancreas, stomach, colon, rectum, breast, uterus, mouth, throat, liver and thyroid.

Zingiber officinale **Linn. (Adrak or Sonth)-** Its rhizome contains camphene, gingerol, zingiberene, borneol, cineol and proteins. It is used in various cancers.

Besides above, some other anticancer dietary (food) plants include *Amorphophallus companulatus* (Suran), *Avena sativa* (Oat), *Cajanus cajan* (Arhar), *Hordeum vulgare* (Jau), *Lens culinaris* (Masur), *Mentha arvensis* (Podina) and *Zea mays* (Makka) (Madhuri and Pandey, 2008; Pandey *et al.,* 2013a).

Antioxidant Vegetables for the Treatment of Cancers

The plant kingdom serves as food and medicinal sources, and thus maintains the health and vitality of human beings as well as animals without causing any toxicity. Medicinal plants, including vegetables are known to have good immunomodulatory antioxidant activities, leading to anticancer effect. They act by stimulating both non-specific and specific immunity, and may promote the host resistance against infection by re-stabilizing body equilibrium and conditioning the body tissues (Madhuri, 2008; Madhuri and Pandey, 2009b). Hence, the consumption of vegetables is widely accepted as lowering the risk of different types of cancer. Vegetables contain several phytochemicals having potent antioxidant activities. The antioxidant vegetables prevent from the cancer by protecting cells from damage caused by 'free radicals'- highly reactive oxygen compounds. Thus, consuming a diet rich in antioxidant vegetables may protect from the occurrence of cancer (Pandey, 2011c; Pandey and Madhuri, 2011a).

Some Antioxidant Vegetables Acting Against Cancer

Some antioxidant vegetables, including spices and oils have been mentioned in Table 11 (Pandey and Madhuri, 2011a). Many doctors recommend that people wish

Table 11: Some Antioxidant Vegetables Acting against Cancer

Botanical (with Hindi or English) Name	Part Used	Main Active Components	Specific Antioxidant/ Anticancer Activity
Abelmoschus esculentus (L.) Moench (Lady's finger)	Fruit, seed	Carotene, vitamins (B, C), amino acids	Seeds inhibit cancer growth
Allium cepa Linn. (Piyaz, onion)	Bulb	Diallyl disulphide, allicin, allin, quercetin antioxidant flavonoid, vitamins (C, E)	Diallyl disulphide inhibits stomachcancer; quercetin may cure lung and other cancers
Allium sativum Linn. (Lasun, garlic)	Bulb	Sulphur compounds (diallyl sulphide, diallyl disulphide, allyl propyl disulphide), allicin	Allicin inhibits cancers of stomach, liver, colon, breast and endometrium; sulphur compounds inhibit cancer cells
Brassica campestris Linn. (Sarson, mustard)	Seed oil	Dithiolethiones, isothiocyanates	Given in tumours/cancers
Brassica oleracea var. *botrytis* Linn. (Cauliflower)	Sprout (flowers)	Ascorbigen, vitamins (A, B, C), sulphoraphane, isothiocyanate	Given in cancers of bladder, lung, stomach, colon, rectum and breast
Brassica oleracea var. *capitata* Linn. (Pattagobhi, cabbage)	Leaf	Ascorbigen, vitamins (A, B, C), sulphoraphane, isothiocyanate	Given in cancers of bladder, lung, stomach, colon, rectum and breast
Brassica rapa Linn. (Shalgam, turnip)	Root, leaf, seed	Ascorbigen, vitamins (A, C), Ca	Given in various tumours/carcinomas
Citrus limon Linn. (Nibu, lemon)	Fruit	Vitamin C, flavonoid, flavone, limonoid, limonene (terpenoid), nobiletin, tangeretin	Flavonoid, tangeretin, nobiletin, limonoid and limonene inhibit cancer growth; nibu inhibits human breast cancer, metastasis and leukemia
Coriandrum sativum Linn. (Dhaniya, coriander)	Leaf, fruit	Essential oil, vitamin C, carotene, borneol, limonene, α-pinene	Antioxidant
Cucumis sativus Linn. (Khira, cucumber)	Fruit, seed	Vitamin C, 14 α-methyl-phytosterol, amyrins	Given in tumours
Curcuma longa Linn. (Haldi, turmeric)	Rhizome (tuber)	Curcumin, β-pinene, camphene, eugenol, curcuminoids, *b* sitosterol	Active against oesophagus, colon, liver, bladder and prostate cancers; given in leukemia, fibrosarcoma and stomach papilloma

Contd...

Table 11–Contd...

Botanical (with Hindi or English) Name	Part Used	Main Active Components	Specific Antioxidant/ Anticancer Activity
Daucus carota Linn. (Gajar, carrot)	Root, leaf	Carotene, flavonoid, carotenoid, glycoside	Given in tumours
Glycine javanica Linn. (Soybean)	Seed	Phytates, saponins, phytosterols, lignans, isoflavonoids, isoflavones	Lowers risk of breast, prostate, colon, stomach, rectum and lung cancers; isoflavonoids and isoflavones inhibit many cancers
Lycopersicon esculentum Mill. (Tamatar, tomato)	Fruit	Vitamins (A, B, C), essential amino acids, lycopene	Antioxidant, anticancer
Mentha spicata Linn. (*M. arvensis*) (Pudina, garden mint)	Whole plant, leaf	Essential oils (menthol, menthone, limonene), flavonoid, sesquiterpene	Active against colon, breast, bladder and prostate cancers; given in lymphoma and leukemia
Momordica charantia Linn. (Karela, bitter gourd)	Leaf, fruit, seed	Vitamins (A, C)	Anticancer
Moringa oleifera Lam. (Munga, Mungana)	Root, leaf	Vitamins (A, C)	Anticancer
Nelumbo nucifera Gaertn. (Kamal, lotus)	Wholeplant	Alkaloids, vitamins, quercetin flavonoid	Antioxidant
Phaseolus vulgaris (French bean)	Fruit	Amino acids, anthocyanin, quercetin	Antioxidant, anticancer
Trigonella foenumgraecum Linn. (Methi, fenugreek)	Leaf, seed	Choline, trigonelline, saponin, amino acids, vitamins, quercetin	Given in various cancers
Zingiber officinale Linn. (Adrak, ginger)	Rhizome	Camphene, gingerol, zingiberene, borneol, cineol, curcumins, proteins	Antioxidant, anticancer; given in tumours

to reduce the risk of cancer must eat several portions of vegetables every day. A study of WHO on diet, nutrition and prevention of chronic diseases recommended that we daily consume at least 400 g of vegetables, including at least 30 g of pulses, nuts and seeds. People who eat much quantity of vegetables have about one-half the risk of cancer and less mortality from cancer. Further, one-third of all cancer deaths in the United States could be avoided through dietary modification, which includes an abundant intake of vegetables and fruits. A large body of scientific evidence indicates that an association exists between inadequate antioxidant status and increased risk for many diseases, including cancer. Antioxidant vegetables have beneficial effect against several diseases. Vegetables contain compounds, such as sulphoraphane that induces GSH transferase, thereby helping detoxifying many carcinogens. Increased consumption of vegetables can increase the plasma antioxidant capacity, and is associated with the lower risk of cancer. Vegetables are most effective against those cancers that involve epithelial cells, *e.g.*, cancers of lung, oesophagus, stomach, colon, pancreas and cervix. The protective effect of vegetables has also been observed for hormone related cancers. The AlE of dhaniya, haldi, karela and adrak were tested for tumour inhibitory effect. Haldi and karela were most effective against metastatic prostate cancer cell lines. Vegetables with the highest anticancer activity are lasun, soybean, pattagobhi, gajar; with a modest level of cancer-protective activity are piyaz, nibu, haldi, phoolgobhi, tamatar; with a low level of anticancer activity is khira (Kaur and Kapoor, 2002; Pandey, 2011c; Pandey and Madhuri, 2011a; Pandey *et al.*, 2013a; Rao *et al.*, 2004; Vecchia and Tawani, 1998).

Some phytochemicals which provide protection against cancer are: allyl sulphides present in lasun and piyaz; glucarates in nibu, began and alu; phytates, lignans, isoflavones and saponins in soybean; isothiocyanates in sarson, phoolgobhi and pattagobhi; and flavonoids, carotenoids and terpenoids in different vegetables. These phytochemicals block various hormone actions and metabolic pathways which are associated with the development of cancer. Soybean is the contributing factor in the low incidence of breast, prostate, stomach, colon, rectum and lung cancers. Soybean seed contains isoflavonoids which inhibit the growth of hormone-dependent and hormone-independent cancer cells in culture. The isoflavones in soy inhibit the growth of human breast and prostate cancers. Nibu due to its flavonoid, tangeretin and nobiletin contents can potentially inhibits the tumour cell growth, and can activate the detoxifying cytochrome P-450 enzyme system. Limonoids present in nibu inhibit the tumour formation by stimulating GST enzyme. Limonene (a terpenoid) content of nibu also possesses anticancer activity. Nibu has been used to inhibit the breast cancer cell proliferation, delay mammary tumorigenesis, and cure metastasis and leukemia cancers. A variety of isoprenoid compounds have been found in the vegetables which show anticancer activities. These compounds include tocotrienols (related to tocopherols) and terpenoids (*e.g.*, limonene, geranoil, menthol and carvone). They increase the tumour latency and decrease the tumour multiplicity. There are many vegetable pigments like flavonoids, carotenoids and anthocyanins which protect from various diseases. Flavonoids extend the activity of vitamin C, act as antioxidant and have antitumour activities. Quercetin (a flavonoid), mainly present in piyaz (both red and yellow) and French bean, possesses anticarcinogenic activity.

The carotenoids are powerful antioxidants that provide protection against oxidative damage, and stimulate immune function. Persons with high levels of serum carotenoids have a reduced risk of cancer. In addition, a variety of phenolic compounds (*e.g.*, caffeic, ellagic, and ferulic acids, sesamol and vanillin) are present in vegetables. These exhibit antioxidant and anticancer activities. Adrak contains phenolic compounds (gingerol), which have antioxidant activity that is even greater than α-tocopherol (vitamin E). Compounds that stimulate the activity of GST are the inhibitors of cancer. Substances that stimulate GST activity are phthalides in celery seeds; sulphides in piyaz and lasun; dithiolethiones and isothiocyanates in sarson, rai, phoolgobhi and pattagobhi; limonoids in nibu; and curcumins in adrak and haldi. Vegetables and fruits also contain a variety of isoprenoid compounds that show anticancer activity. These compounds include tocotrienols (related to tocopherols) and terpenoids (*e.g.*, limonene, geranoil, menthol and carvone). Thus, they increase the tumour latency and decrease tumour multiplicity (Heber, 2004; Pandey, 2011c; Pandey and Madhuri, 2011a; Pandey *et al.*, 2013a; Vecchia and Tawani, 1998).

Phytoremedies of Malnutritional Cancers in Animal

In the so called *'Third World'*, where poverty is common, millions of poor people are engaged in hard labours. Millions and millions of animals like horses, donkeys, bullocks and he-buffaloes, etc. work day in day out in cart or for other carriage. All of them don't get good food and don't get enough to drink. They have to work in the burning sun, on busy and dusty roads, without one moment of rest. Next to that, some individuals don't even know that *'What is pain'*. A lot of poor people and different species of animals have to work till they drop dead, and that often happens in the streets. They are day-to-day exposed to numerous environmental hazards and disasters, including xenobiotics, contaminants, radiations, genetic factors, etc. All these result in *'malnutrition'* which may cause *'cancer'*, i.e., *'malnutritional cancer'*. Overall, the poverty and decreased dietary intake of good quality are the most important causes of malnutritional cancer. The animals like horses, donkeys, bullocks and he-buffaloes, etc. work day in day out in cart or for other carriage. All of them don't get good food and don't get enough to drink. They have to work in the burning sun, on busy and dusty roads, without one moment of rest. Now, cancer is a burning problem in this westernized *'ultra-modern era'* even in our country (Pandey, 2014; Pandey and Sahni, 2011a).

Eating too little or too much will result in poor health. The food eaten must not only be nutritious but it must also be complete and clean, otherwise the individual eating the food would get ill even the food is nutritious. So, if the right food is not consumed in right amounts, it results in *'malnutrition'* (*mal-noo-trish-un*) leading to either *'undernutrition'* or *'overnutrition'*. Malnutrition occurs when we do not get enough calories or nutrients and hence, it may cause several diseases, including cancer. Malnutrition acts as carcinogen in many ways to produce cancer in human beings and animals, and so the aetiological role of malnutrition preceding clinical cancer has been firmly established. In India, malnutrition is the single largest cause of the mortality and morbidity not only in infants, children and working or pregnant women, but also in livestock animals. The single major factor responsible for the wide prevalence of malnutrition in India is the poverty. Like many countries, India is also facing a great problem of malnutrition. Thus, similar to the human beings, the animals (*viz.*, horse, cattle, buffalo, sheep, goat, pig, dog, poultry, and also wild animals, etc.) suffering from malnutrition need special attention. Those in moderate and mild degree of malnutrition require additional food in order to bring their weight to normal; whereas, those suffering from a severe degree of malnutrition need special attention by the owner as well as veterinarian because such animals usually suffer from diarrhoea, respiratory infections and other diseases. It has been clearly stated that malnutrition predisposes to infection, or infections lead to malnutrition. The body requires energy to carry out the different activities, and proteins are necessary for growth and repair. Deficiency of these two basic nutrients in our body leads to *'protein energy malnutrition'* (PEM). This disorder has very serious consequences, leading to cancer and/or death (Pandey, 2007, 2010 and 2014).

From ancient times we, in India, have endowed food with magical qualities. In the Vedic times, food was associated with divine attributes. This tradition was prevalent in other societies as well- ancient Egypt, for example. Ancient people discovered the healing powers of food- perhaps initially by accident. Later, these patterns became well established and were described in medical and veterinary texts. It is quite evident that the diseases which can be cured by food or nutrient concentrates are caused by deficiency of particular nutrients. Thus, the malnutritional disorders can be cured by giving food sources rich in the nutrient that is otherwise lacking in the diet, and there is no doubt that food can help to control several ailments. Treatment of malnutrition may include increasing the calories and nutrients in diet. If malnutrition results in cancer, it is very difficult to treat, but some improvements can be brought about by proper foods and drugs. Episodes in the history of man's battle against disease clearly explain that food can cure many diseases. Thus, food or *'phytotherapy or phytoremedy'* can be the best remedy for malnutritional cancer (Pandey, 2007 and 2014; Pandey and Sahni, 2011a).

Malnutritional Cancers in Animals

If an individual eats the right kind of foods in the required amounts, that individual will keep good health provided no other factors intervene. On the other hand, a poor eating pattern or eating too little or too much will result in poor health. These are both facets of malnutrition. When the diet supplies too little of one or more nutrients, a kind of malnutrition called *'undernutrition'* occurs. When the diet provides too much of one or more nutrients, another form of malnutrition called *'overnutrition'* results. Malnutrition most often refers to undernutrition resulting from an inadequate consumption, poor absorption, or excessive loss of nutrients, but the term can also encompass overnutrition resulting from excessive intake of specific nutrients. An individual will experience malnutrition if the appropriate amount of, or quality of nutrients comprising a healthy diet are not consumed. An extended period of malnutrition can result in starvation or many other diseases such as scurvy and cancer. The individual suffering from malnutrition almost always have infections such as diarrhoea and respiratory infections. Just as infections are common in cases of severe malnutrition, infections like diarrhoea and whooping cough can also cause malnutrition. Thus, malnutrition can increase the risk of infections and infections can, in turn, lead to malnutrition. This inter-relationship and synergistic effect of malnutrition and infections often lead to a high incidence of cancer and/or deaths. Malnutrition leads to infection by reducing the diseases-fighting capacity of the body. Many environmental or pollution factors, including poverty contribute to eating disorder or malnutrition (Pandey, 2007 and 2014; Pandey and Madhuri, 2010e).

In animals, PEM may result due to deficiency of energy and protein. This disorder is of three types, namely: (a) Marasmus (deficiency of both carbohydrate and protein); (b) Kwashiorkor (deficiency of only protein); and (c) an intermediate state of marasmus and kwashiorkor. *'Marasmus'* is characterized by very low body weight and weight loss due to growth failure, stunting, loss of subcutaneous fat, gross muscle wasting, emaciation and failure to thrive. In addition, watery diarrhoea is associated often with dehydration, irritation (doesn't allow to touch), decreased responsiveness, behavioural changes, no oedema and wrinkled dry skin dry. Other deficiencies particularly, vitamin A deficiency may occur. *'Kwashiorkor'* is characterized by oedema of whole body (especially belly), wasted muscles and retarded growth. Other signs and symptoms include psychomotor changes, hepatomegaly (enlarged, fatty liver), irritability, skin changes (flaky paint appearance and mosaic skin appearance), vitamins A and B deficiencies and associated diseases such as watery diarrhoea, severe respiratory infection (cough), etc. In intermediate type of PEM, the features of both marasmus and kwashiorkor PEM are seen. Associated symptoms like vitamin C deficiency, deficiency of Zn, atrophy of papillae on the tongue, angular stomatitis, xerophthalmia, etc. may occur. PEM can result in marked hypoalbuminemia, anaemia, oedema, muscle atrophy, delayed wound healing, impaired immunocompetence, weight loss, reduced basal metabolism, depletion of subcutaneous fat and tissue, bradycardia and hypothermia. The patient with PEM is typically emaciated, elderly and chronically ill. Thus, malnutrition places stressed patients at a greatly increased risk for morbidity, leading to cancer and mortality. Moreover, there may be a longer recovery period, impaired host defenses and sepsis, impaired wound healing, anemia, impaired gastrointestinal tract function, muscle atrophy, impaired cardiac function, impaired respiratory function, reduced renal function, brain dysfunction, delayed bone callus formation and atrophic skin (Pandey, 2007 and 2014).

In animals, the common environmental factors leading to cancer may be bad feed or diet, infections, radiation, hard working and environmental pollutants. These factors cause abnormalities in the genetic material of cells. Cell reproduction is an extremely complex process, which is normally tightly regulated by several classes of genes, including oncogenes and tumour suppressor genes. Hereditary or acquired abnormalities in these regulatory genes can lead to developmemt of cancer. Abnormal diets such as fad or limited diet and recent decrease in food intake usually cause malnutrition. There is an increased risk of malnutrition associated with chronic diseases such as cancer, and diseases of intestinal tract, kidney and liver. Patients with these chronic diseases, especially cancer may lose weight rapidly. Studies show that malnutrition is closely linked to the income level of the owner of animal. The animals given with high drug doses are also at increased risk of malnutritional cancer. They tend to maintain inadequate diets for long periods and their ability to absorb nutrients is impaired by the drug's effect on body tissues, particularly the liver, pancreas and brain. Malnutrition usually develops in stages over a long period of time. Some early signs and symptoms may include: irritable (bad mood) and tired, slower growth than normal, or no growth, and weight loss. Bone or joint pain, weak muscles, bloated abdomen, swelling in other parts of the body, change in the colour of skin and hair, hair loss, loss of appetite, slow wound healing, and easily getting infections are noticed. All these signs and symptoms are correlated with those of many cancers. Animals with chronic diseases, especially cancer may lose weight rapidly and become susceptible to undernourishment because they can not absorb valuable vitamins, calories and iron (Pandey, 2014).

In relation to cancer, malnutrition is the carcinogenic effect produced by nutritional variables through multiple endogenous involvements. Nutritional factors act as primary effectors in four situations: carcinogens in food articles, affected bioavailability of nutrients, non-nutritive dietary items and harmful contaminants. Nutritional carcinogenesis may occur due to ingestion of toxins, malnutrition, non-bioavailability of the micronutrients and inactivation of the metabolic enzymes (*i.e.*, mixed-function oxidases) present in the liver. Thus the aetiological role of malnutrition preceding clinical cancer has been firmly established. There are certain environmental substances (contaminants or hazards) which can cause malnutritional cancer both in humans and animals. Most of the environmental hazards have been shown as carcinogenic, or they may contaminate the diet or food, thereby causing malnutrition. Many factors have a major effect on increasing the rates of oral, colon, lung and mammary gland cancers. Some of these factors include increased infections, more use of pesticides, low consumption of fruits and vegetables, high industrial pollution, more exposure to sun, decreased physical activity, etc. Almost every fruit, vegetable, etc. contains natural carcinogenic pesticides (Pandey and Madhuri, 2010e; Pandey and Sahni, 2011a).

Malnutrition is very common in sick animals, and is thought to affect up to 50 per cent of hospitalized dogs and cats. Within days, malnutrition can begin to have deleterious effects on immune function, healing and, most importantly, quality of life. While human patients will follow their doctor's instruction to eat even though they do not feel well or have a good appetite, it is impossible to convince a sick pet to eat if they do not feel like it. Due to malnutrition, different types of cancer may develop in dogs. Some symptoms of dog cancer are rather indistinct and, especially at the onset of the disease, it is difficult to tell whether a dog, showing some cancer-like symptoms, is suffering from canine cancer or some other illnesses. Some cancers (*e.g.*, lymphoma, mast cell tumour and fibrosarcoma) can arise as lumps and bumps on the dog's skin. Very often, benign growths such as warts, fatty tumours, cysts, etc. can arise in a dog's skin as well. Usually, any lumps and bumps that appear, then decrease in size or even disappear, and finally reappear in a larger size are more likely to be cancer than benign lumps. Coughing and sneezing are common symptoms of dog cancer. Chronic disorders of gastrointestinal tract (GIT), such as vomiting and diarrhoea can be symptoms GIT cancer of dog, or cancer anywhere else. Appetite loss is also one of those symptoms of dog cancer that is indistinct, as many other diseases (*e.g.*, dental or gum diseases, inflammatory bowel disease, heart disease, kidney disease, etc.) can cause appetite loss in dogs as well. Also, older dogs tend to eat less. Chronic skin diseases such as skin sores, itch and lesions, and wounds that do not heal, can be indicative of canine skin cancer. Blood in urine or urinary incontinence can be caused by some types of dog cancer (*e.g.*, bladder cancer), but they can also be caused by urinary tract infections, bladder stones, kidney disease, diabetes, Cushing's disease, etc. While cancer can cause a lot of pain to a dog, so can many other diseases, such as arthritis, ear infections, dental or gum diseases, etc. Bone cancer in dogs can cause lameness. However, lameness can also be caused by arthritis, hip dysplasia, or some form of injuries. Perhaps lethargy and weakness are one of the most vague symptoms of all diseases. A dog can be lethargic due to a various reasons, *e.g.*, pain from arthritis, anaemia, fever caused by infections, etc. Oral cancer in dogs can cause bad breath, but perhaps more frequently, bad breath is caused by indigestion, dental/gum diseases, liver disease and kidney disease. Brain tumours in dogs can cause seizure, but it can also be seen in thyroid problems, kidney disease, liver disease, hypoglycemia and poisoning. Many of the above symptoms of dog cancer are not unique to cancer. That is why very often cancer is not diagnosed until it is at a later stage. Finally, whenever a dog has a disorder that does not respond well to treatment in one to two months, and is showing some symptoms of dog cancer, then the possibility of cancer should be considered and a more thorough examination should be done (Pandey, 2014).

The owners of horses and donkeys are often very poor. The owner really needs the money, because he has to take care of his family. When the animal becomes sick, the owner is not able to make money because he does not have the horse or donkey to pull. In malnutrition, the horses grow out of shape. Because of all the heavy works, the horses can not develop physically; often legs and ankles become crooked. They have often a bad back also, since they have to carry a too heavy load and because it is loaded in a wrong manner. The animals have to work in the burning sun with terribly high temperature. The owners do not give them enough to drink, so they dehydrate. Due to less consumption of water and food, the animals are exhausted and overly tired. Through malnutrition, starvation, dehydration and exhaustion occur. In the countries like Netherland, Egypt, Jordan, Pakistan and India, the cars are expensive, so horses pull carriages that transport people; and the horses and donkeys have to pull carts with heavy loads, *e.g.*, building materials. These animals are seen as *'motor'* to do the heavy works. These countries are usually very hot and we can not cope with that, but the horses and donkeys have a problem with it also. In the World War I (1914-1918), the British army was present in Egypt. In the beginning of this war, about 20,000 horses and donkeys from the United Kingdom were brought ashore. When the war ended in 1918, the horses and donkeys were sold to Egyptian people, so the poor animals were left there (Pandey, 2014).

Edible Plants in Phytoremedy of Malnutritional Cancers

Treatment for malnutrition depends on the cause. Vitamin or mineral supplements are needed if the body is low in vitamins and minerals. Medicines may also be given if health problem appears due to malnutrition. Diet therapy helps patients to lead a full life. To prevent energy deficiency, carbohydrate-rich foods (*e.g.*, sugars, cereals, roots and tubers, and fruits like banana, sapota and mango) and fat-rich foods (*e.g.*, nuts, oilseeds, vegetable oil, ghee, vanaspati and butter) should be taken. If protein is lacking in diet, pulses, nuts and oilseeds should be given. If vitamins and minerals are deficient, green leafy vegetables, nuts, pulses, etc. are to be given. In addition, fibre should also be eaten to prevent diseases like constipation and cancer of colon. Some complex carbohydrates (*e.g.*, cellulose) and non-carbohydrate forms of fibre are available, which may be helpful to treat colon cancer. In PEM, high energy foods in case of mild to moderate diarrhoea, a high energy-high protein diet when tolerance improves for solid foods, and mineral and vitamin supplements, etc. are given. In chronic diseases like cancer in which tissue wasting takes place, a high caloric-high protein diet is the best. Herbivorous animals with malnutritional cancer must be given the feed containing sufficient quantity of fruits and vegetables. It is suggested that vegetables and fruits have strong protective effect against major diseases, including cancer (Pandey, 2007 and 2014; Pandey and Madhuri, 2010e).

Many naturally occurring substances present in the diet/food have been identified as potential chemopreventive agents. Various agriculture plants and their products have been used traditionally for the treatment of various livestock diseases since times immemorial. The dependence of rural mass on the plant-based medicines for curing animals is mainly because of the limited access to the modern (allopathic) medicine system, cheaper and easy availability, and the simplicity of their applications. The herbal drugs also do not possess toxicity or have very less side effects (Jadeja *et al.*, 2006). The edible plants are used not only for food but also employed to cure different ailments of livestock. The medicinal uses of such plants may help veterinarians, medicos and agriculture/bio scientists or researchers in discovering new drugs against various diseases of human and animals (Pandey and Madhuri, 2008c).

Many dietary agricultural plants and their products are commonly consumed by humans as well as animals as food matters. They prevent from malnutritional cancer and various other diseases. Some of the dietary plants which have been reported as *'phytotherapy/phytoremedy'* for malnutrition and cancer are described below (Pandey, 2007 and 2014; Pandey and Sahni, 2011a).

The bulb of *Allium cepa* (Piyaz) contain diallyl disulphide, quercetin flavonoid, allicin, allin, amino acids and vitamins (folic acid, C and E). Diallyl disulphide inhibits cancer cells in stomach. Quercetin, because of its antioxidant property, treats lung cancer and some other cancers. The bulb of *A. cepa* var. *aggregatum* (shallot) contains allyl propyl disulphide, flavonoid, phenolic content, proteins and vitamins (folic acid and C). Allyl propyl disulphide, phenolic content and flavonoid inhibits cancer cells in liver, stomach and other organs. It also eliminates toxins to inhibit and kill the cancer cells in liver. *A. sativum* (Lasun) bulb contains sulphur compounds (diallyl sulphide, diallyl disulphide, allyl propyl disulphide), allicin, allin and amino acids. The sulphur compounds inhibit cell proliferation, modulate cell cycle activity and interfere with hormone action in cancer cells. Allicin inhibits the proliferation of cancer cells of mammary gland, endometrium and colon. Lasun is also used in gastric and liver cancers. In animals, the bulbs are also given in malnutritional disorders of skin and lung, flatulence, and dyspepsia. *Annona squamosa* (Sitaphal) contains annoaine, higenamine and amino acids. Its leaves and seeds are externally applied over malnutritional wounds, boils and swelling in animals. The riped fruit (Sitaphal) is rich of carbohydrate (energy), so it can be given to cure marasmus PEM. *Brassica campestris* (Sarson) seed oil contains glycerides of palmitic, stearic, oleic and linoleic acids. It is used in various tumours. In animals, the oil is also given in rheumatism, cancer and ulcer. *Br. juncea* (*Br. nigra,* Rai) plant contains essential oils, while its seed contains two antithiamines and flavonol glycosides. In animals, oil along with rock salt is used in gum disease. Seeds and oil of Rai are used in malnutritional rheumatism, paralysis and skin diseases. The sprouts (flowers) of *Br. oleracea* var. *botrytis* (Phoolgobhi) contain cysteine, ascorbigen, sulphoxide, indole-3-carbinol, glucaric acid, sulphoraphane, glucosinolates, isothiocyanates and vitamins (A, C). These components possess anticancer activities against lung, stomach, colon, bladder, mammary and rectum cancers. Its leaves contain vitamin C, and are externally used in gout, rheumatism, skin diseases and blister in animals; while phoolgobhi is cardiotonic and antiinflammatory, and may be used in biliousness and urinary diseases. *Br. oleracea* var. *capitata* (Pattagobhi, cabbage) contains glucaric acid, sulphoraphane, glucosinolates, isothiocyanates, cysteine, ascorbigen, sulphoxide, indole-3-carbinol, allyl isothiocyanate and vitamins (A, B, C). These have anticancer activities against bladder, lung, stomach, colon, rectum and mammary cancers. *Br. rapa* (Shalgam) leaves and roots contain ascorbigen and vitamins (A, C), and are used in various tumours. *Cajanus cajan* (Arhar) leaves contain cajanol, while roots contain cajanone and cajaquinone. Its seeds are quite rich in proteins, so the seeds (or cooked 'dal') can be given in marasmus and kwashiorkor PEM. The expressed juice of leaves, with salt is given in animals to cure malnutritional jaundice.

The fruits of *Capsicum annum* (Lalmirch) contain amino, ascorbic and folic acids, glutathione, and flavonoids. In animals, the fruit powder with oil cures foot diseases. Fruits are stimulant and stomachic, and are used in dyspepsia, neuralgia, rheumatism and diarrhoea. *Carica papaya* (Papita) fruit contains papin and cryptoflavin, while the seeds contain carpasemine. In animals, the seeds and fruits are given to cure hepatitis, diarrhoea and dysentery. The fruits are digestive, stomachic, carminative and diuretic too. *Cicer arietinum* (Chana) seed contains isoliquiritigenin, while the plant contains amino acids and vitamins (A, D, E). Its fresh leaves juice is given to livestock in constipation. The macerated seeds in water are used as tonic. The seeds are astringent, diuretic and antibilious, and are given in debility, scurvy and jaundice. Chana seeds (or cooked *'Dal'*) are quite rich in proteins, so the seeds (or cooked *'Dal'*) can be very useful in marasmus and kwashiorkor PEM diseases. *Citrus limon* (Baranibu) fruit contains flavonoid, flavone, limonoid, limonene, nobiletin, tangeretin and vitamin C. The flavonoid, tangeretin and nobiletin are potent inhibitors of tumour cell growth, and can activate the detoxifying P-450 enzyme system. Limonoids inhibit tumour formation by stimulating the enzyme GST. The limonene (a terpenoid) also possesses anticancer activity. Nibu fruit is used for inhibition of mammary cancer cell proliferation and delaying of mammary tumorigenesis. It is also used in metastasis and leukemia. The Leaves of *C. limon* contain bergapten, citropten and limettin. The fruits are antiseptic, appetizing, astringent and stomachic, and are given in scurvy, rheumatism, dysentery, diarrhoea and liver disorders of animals. *Curcuma longa*

(Haldi) rhizome contains curcumin, zingiberine and curcuminoids. Haldi acts against colon, bladder and prostate cancers, intravesical tumour, fibrosarcoma, HCC, oesophagal carcinogenesis, leukemia, stomach papilloma and solid tumour cell lines. In large animals, the rhizome powder is used in pain, swelling, wound, mastitis, tumour and liver diseases. *Daucus sativa* (Gajar) root contains pyruvate kinase, acyltransferase and alkane; while seeds contain sitosterol and amino acids. The boiled roots are given to cattle to fatten them. The roots and seeds are stimulant and carminative, and are used in dropsy and kidney diseases. In animals, the decoction of roots and leaves of *Elephantopus scaber* (Gobhi) are given in dysuria, diarrhoea, dysentery, swelling or pain in stomach, ulcer and eczema. Its whole plant is astringent and cardiac tonic. *Emblica officinalis* (Amla) fruit contains ascorbic acid (vitamin C) and phyllembic acid. It inhibits cellular mutation and may prevent cancer. Its 18 compounds inhibit the growth of gastric and uterine cancer cells. Amla also inhibits the growth of *in vitro* breast cancer cells. *Glycine javanica* (Soybean) seed contains phytates, protease inhibitors, phytosterols, saponins, isoflavonoids and isoflavones. The soybean is contributing factor in the low incidence of breast, prostate, stomach, colon, rectum and lung cancers. Its isoflavonoids inhibit the growth of hormone-dependent and hormone-independent cancer cells in culture; however, its isoflavones inhibit the growth of mammary and prostate cancers. The soy seeds (or cooked 'Dal') are quite rich in proteins, so the seeds (vegetables) may potentially treat both marasmus and kwashiorkor PEM. *Lens culinaris* (Masur) plant contains lenticin, tricetin and luteolin; while seed contains protein. In large animals, masur gruel is given in haematuria, and paste of seeds is applied on the wounds and broken horns. The seeds are laxative and given in constipation. Since the seeds of *L. culinaris* are rich in protein, these (as Masur Dal) can be used to cure both marasmus and kwashiorkor PEM diseases.

Lycopersicon esculentum (Tamatar, tomato) fruits contain lycopene, which acts as an antioxidant and inhibits the proliferation of cancer cells. Higher intake of tomato is correlated with protection from the cancer; the protective effect is strongest for cancers of prostate, lung and stomach but some protective effects also appear for cancers of pancreas, colon, rectum, esophagus (throat), mouth, mammary gland and cervix. Tomato fruits also contain oxalic, citric and mallic acids, and vitamins (A, C); while its seeds and herbage contain solanine. Its pulp and juice are intestinal antiseptic, blood purifier, cholagogue, stimulant and digestive, and are given in pulmonary diseases and dyspepsia of animals. *Momordica charantia* (Karela) leaf, fruit and seed contain linoleic acid, momordin, palmitic acid, proteolytic enzymes and vitamins (A, B, C). These parts are used in colon, mammary and bladder carcinomas, lymphoma, leukemia and other tumours. Fruits and seeds are carminative and stomachic, and are prescribed in rheumatism, gout and liver diseases of animals. The silver skin of *Oryza sativa* (Chaaval) contains oryzanin; while bran contains a glucoside nukain, which yields aglucone nukagenin. In animals, Chaaval with Dal is used as a perfect food in dysentery and can alone support life as it contains protein, carbohydrate, fat, vitamins and minerals. Thus, the Chaaval may cure both marasmus and kwashiorkor PEM diseases. Decoction of grains is used internally as antiinflammatory for digestive tract.

Picrorhiza kurroa (Kutki) fruits/seeds contain kutkin and kurrin. In animals, the fruits are given in loss of appetite, dropsy, fever, and foot and mouth disease (FMD). The roots are bitter, stomachic and purgative. *Psidium guajava* (Amrud) plant contains leucocyanidin and quercetin; while the leaves contain eugenol. In animals, the decoction of root bark and leaves is given in chronic diarrhoea. Its root bark is astringent and fruits are laxative; while the leaves are astringent, and applied in wounds and ulcers. *Raphanus sativus* (Muli) root contains glucoside, enzymes and methyl mercaptan. In animals, the roots are used in dysentery, indigestion, urinary disorders and gastrointestinal pain. Its seeds are diuretic, laxative and carminative. The seeds of *Sesamum orientale* Linn. (Til) contain essential oil, proteins sesamin and sesamolin, so the seeds and oils can be given in PEM diseases of animals. *Solanum melongena* (Baingan) fruit contains vitamins A, B and C; while seed contains linoleic and oleic acids. In animals, the fruits and leaves are given in weakness, dysuria, liver disorders and ulcers. The seeds of *S. melongena* are stimulant. *Solanum tuberosum* (Alu) plant contains solanine, solasodine and lutein. In livestock, a portion of raw tuber (Alu) is grated over the sore to cure leg ulcers. Alu is full of carbohydrate; it can be given in weakness, particularly in PEM. *Spinacia oleracea* (Palak) plant contains vitamins, iodine, lecithin, carotene and amino acids. The whole plant and leaves of *S. oleracea* are used in inflammation of lung, liver and bowel, fever and jaundice in animals. Its seeds and fruits are laxative, demulcent and diuretic. *Trachyspermum ammi* (Ajwain) fruits yield essential oil containing thymol. In animals, the fruits are given in weakness after pregnancy and low lactation. They are also employed in gastric disorders and diarrhoea, and have antiseptic, carminative and stimulant properties. *Trigonella foenumgraecum* (Methi) leaves and seeds contain choline, trigonelline, amino acids, proteins, vitamins and quercetin. These parts are used in various cancers. In animals, its leaves and seeds are used in rheumatism, cancer, intestinal inflammation, indigestion and bilious disorders. Its seeds are carminative and tonic too. *Triticum aestivum* (Gehun) grains contain ellagic, linoleic and oleic acids, oligosaccharides, fat, sterols, phytases, tocotrienols and vitamin E. Its whole grains protect against lymphomas and cancers of pancreas, stomach, colon, rectum, breast, uterus, mouth, throat, liver and thyroid. In animals, the flour of wheat is given in weakness and gastric disorders.

Zea mays (Makka) grains contain oxalic acid, carbohydrate, fat, protein and vitamins. Its grains are astringent, nutritive and nourishing in humans and animals. *Zingiber officinale* rhizome (called 'Adrak' and dried rhizome called 'Sonth') contains a volatile oil containing camphene, gingerol, zingiberene, borneol, cineol and proteins. It is used in various cancers. In livestock animals, the rhizomes are given with salt before feed to increase appetite and in weakness after pregnancy. They are also given in swelling, indigestion, flatulent colic, and are stimulant and carminative. Some other anticancer dietary plants include *Amorphophallus companulatus* (Suran), *Avena sativa* (Oat), *Cajanus cajan* (Arhar), *Hordeum vulgare* (Jau), *Lens culinaris* (Masur) and *Mentha arvensis* (Podina).

Fruits and Vegetables for Treatment of Malnutritional Cancer

Nutrition is closely associated with health. If a person eats right kind of foods in right amounts, that person will keep good health provided no other factors intervene.

An individual will experience malnutrition if the appropriate amount of, or quality of nutrients are not consumed for an extended period of time, which can result in starvation or many other diseases such as scurvy and cancer. Malnutrition is directly responsible for 300,000 deaths per year in children younger than 5 years in developing countries and contributes indirectly to over half the deaths in childhood worldwide. Globally, with 113.4 million children younger than 5 years affected as measured by low weight for age. The overwhelming majority of these children, 112.8 million, will live in developing countries with 70 per cent of these children in Asia, particularly the south-central region, and 26 per cent in Africa. An additional 165 million (29 per cent) children will have stunted length/height secondary to poor nutrition. Although eating disorders (malnutrition) are treatable and many people recovered, but the recovery is a complex process that can take several months or even years (Pandey, 2007 and 2014; Pandey and Madhuri, 2010e).

Malnutritional Cancer

If the right food is not consumed in right amounts by a person it results in malnutrition, which may cause cancer. Hence, malnutrition is responsible for the development of *'malnutritional cancer'* as a consequence of which a huge mortality occurs every year in human beings, especially in children all over the world. If malnutrition results in cancer, it is very difficult to treat, but some improvements can be brought about by taking proper foods or nutrients and drugs. In spite of many allopathic treatments (*e.g.,* anticancer drugs or chemotherapy, radiation therapy, etc.), no relief or improvements occur. Hence, the individuals should be advised to take sufficient nutritious diets, including more amounts of various fruits and vegetables. The protective and curative actions of fruits and vegetables against cancer have been attributed on the ground of several scientific reports (Pandey and Madhuri, 2010e).

India is facing a great problem of malnutrition. This condition is not a new health problem in India. As far back as 1946, when the Bhore Committee reviewed the health status of the Indian population, diet surveys had already been done. These surveys showed that about 30 per cent of the families surveyed consumed inadequate amounts of food to provide the necessary nutritional requirements. At that time, it was also known that an adequate diet contributed to high mortality and morbidity in the general population, particularly in infants and pregnant women. The single major factor responsible for malnutrition in India is the poverty. About 50 per cent of our people live below poverty line and even after spending 80 per cent of their income on food, they can not have a balanced diet. To combat malnutrition, various supplementary nutrition programmes have evolved as short-term measures (Pandey, 2007 and 2014; Pandey and Madhuri, 2010e).

Malnutrition predisposes a person to infection. On the other hand, infections lead to malnutrition. The people who are at greater risk of having malnutrition include infants and children who do not get the right type or amount of calories and nutrients; older adults; people who are taking medicines that decrease appetite or affect the digestion and absorption of nutrients; people who abuse drugs or alcohol; people with eating disorders; people with certain diseases such as cystic fibrosis, celiac disease, liver disease, kidney disease and cancer; people with a low income who have trouble to buy the right kind of food; and pregnant women. Hence, malnutrition may cause several diseases, including cancer. It has been further stated that the abnormal diets such as fad or limited diet and recent decrease in food intake usually cause malnutrition, which may lead to the development of cancer. Poverty and lack of food are the primary reasons why malnutrition occurs in the United States. Ten per cent of all the members of low income households do not always have enough healthful food to eat, and malnutrition affects one in four elderly Americans. There is an increased risk of malnutrition associated with chronic diseases such as cancer, and diseases of intestinal tract, kidney and liver. Patients with these chronic diseases, especially cancer may lose weight rapidly and become susceptible to undernourishment because they can not absorb valuable vitamins, calories and iron (Pandey, 2007 7 2014; Pandey and Madhuri, 2010e).

People with drug or alcohol dependencies are also at increased risk of malnutrition. These people tend to maintain inadequate diets for long periods and their ability to absorb nutrients is impaired by the alcohol or drug's effect on body tissues, particularly the liver, pancreas and brain. Unintentionally losing 10 lb (4.5 kg) or more weight may be a sign of malnutrition. People who are malnourished may be skinny or bloated. Their skin is pale, thick, dry, and bruises easily. Rashes and changes in pigmentation are common. Hair is thin, tightly curled and pulls out easily. Joints ache and bones are soft and tender. The gums bleed, and tongue may be swollen or shriveled and cracked. Visual disturbances include night blindness and increased sensitivity to light and glare. Anaemia, diarrhoea, disorientation, goiter, loss of reflexes, lack of coordination, muscle twitching, and scaling and cracking of lips and mouth are other symptoms. All these signs and symptoms are correlated with those of many cancers (Pandey, 2007; Pandey and Madhuri, 2010e).

Role of Fruits and Vegetables in Malnutritional Cancer

The poor diets can cause diseases, or some diseases can be cured or controlled by particular diets. From ancient times, India has endowed food with magical qualities. In the Vedic times, food was associated with divine attributes. This tradition was prevalent in other societies as well-ancient Egypt, for example. Ancient people discovered the healing powers of food- perhaps initially by accident; later, these patterns became well established. Another episode in man's search for a cure is the story of scurvy. The scurvy is caused by vitamin C deficiency. Fresh herbs, oranges and other citrus fruits were found to be effective in preventing and controlling scurvy. These episodes in the history of man's battle against disease clearly explain the fact that food can cure certain diseases. Thus, there is no doubt that food can help to control many ailments (Pandey, 2007).

To prevent energy deficiency, carbohydrate-rich vegetable sources like cereals, roots and tubers, fruits like as banana, sapota and mango, and fat-rich foods like nuts, oilseeds, vegetable oil, ghee and vanaspati should be consumed. If protein is lacking in the diet, vegetable foods such as pulses, nuts and oilseeds should be taken. If the vitamins and minerals are deficient in the diet, green leafy vegetables, nuts, pulses, etc. are to be eaten. In addition, fiber can also be given in the prevention of colon cancer. In cancer in which tissue wasting takes place, a high caloric-high protein diet is best. For this purpose, a suitable dietary guideline follows as: cereals-400 g, pulses- 50 g, roots and tubers- 100 g, green leafy vegetable- 200 g and other vegetables- 200 g (Pandey, 2007 and 2014).

Fruits and vegetables play an important role in the treatment and prevention of cancer. Consumption of large amount of fruits and vegetables can prevent the development of cancer. An inverse relationship has been suggested between the consumption of fruits and vegetables and the incidence of cancer in multiple organs including lung, larynx, mouth, pharynx, GIT and pancreas. The intake of 400 to 600 g per day of vegetables and fruits can reduce the occurrence of many common forms of cancer, and diets rich in plant foods can also lower the risk of heart disease and many chronic diseases. Further, one-third of all cancer deaths in the United States could be avoided through dietary modification, which includes an abundant intake of fruits and vegetables. People who eat much quantity of vegetables have about one-half the risk of cancer and less mortality due to cancer. The protective action of fruits and vegetables against cancer has been attributed to the presence of antioxidants, especially antioxidant vitamins. Fruits and vegetables contain several phytochemicals which possess strong antioxidant activities. Thus the fruits and vegetables prevent from cancer and other diseases by protecting cells from damage caused by oxygen free radicals (Pandey, 2007 and 2014; Pandey and Madhuri, 2010e).

Fruits and vegetables contain compounds like sulphoraphane that induces GSH transferase, thereby helping detoxifying many types of carcinogens. Increased consumption of vegetables can increase the plasma antioxidant capacity and is associated with the lower risk of cancer. The vegetables and fruits are most effective against those cancers that involve epithelial cells such as cancers of lung, oesophagus, stomach, colon, pancreas and cervix. The protective effect of vegetables has also been observed for hormone related cancers. The alcoholic extracts of dhaniya, haldi, karela and adrak have been tested for tumour growth inhibitory effects. Haldi and karela were most effective against metastatic prostate cancer cell lines. Vegetables with the highest anticancer activity included lasun, soybean, pattagobhi, gajar; with a modest level of cancer-protective activity include piyaz, nibu, haldi, phoolgobhi, tamatar; with a low level of anticancer activity included khira. The pigments present in the fruits and vegetables, such as flavonoids, carotenoids and anthocyanins protect from various diseases. Quercetin chiefly present in red and yellow piyaz and French bean possesses both anticarcinogenic activity. The carotenoids are powerful antioxidants that provide protection against oxidative damage, and stimulate immune function. Persons with high levels of serum carotenoids have a reduced risk of cancer. A variety of phenolic compounds (caffeic, ellagic, and ferulic acids, sesamol and vanillin) are present in fruits and vegetables. They exhibit antioxidant and anticancer activities. Adrak contains gingerol, which has antioxidant activity that is even greater than α-tocopherol (vitamin E). Compounds that stimulate the activity of GST are considered as inhibitors of cancer. Substances that stimulate GST activity include phthalides in celery seeds; sulphides in piyaz and lasun; dithiolethiones and isothiocyanates in sarson, rai, phoolgobhi and pattagobhi; limonoids in nibu; and curcumins in adrak and haldi (Pandey and Madhuri, 2010e).

The phytochemicals like allyl sulphides present in lasun and piyaz; glucarates in nibu, began and alu; phytates, lignans, isoflavones and saponins in soybean; isothiocyanates in sarson, phoolgobhi and pattagobhi; and flavonoids, carotenoids and terpenoids in other fruits and vegetables provide protection against cancer. They block various hormone actions and metabolic pathways associated with cancer. The isoflavones present in soybean is the contributing factor in the low incidence of breast, prostate, stomach, colon, rectum and lung cancers. Nibu fruit due to its flavonoid, tangeretin and nobiletin contents can potentially inhibit the tumour cell growth and can activate the detoxifying cytochrome P-450 enzyme system. Limonoids present in nibu inhibit tumour formation by stimulating the enzyme GST. The limonene content of nibu also possesses anticancer activity. Nibu is used to inhibit the breast cancer cell proliferation, delaying of mammary tumorigenesis and curing the metastasis and leukemia cancers. Vegetables and fruits also contain a variety of isoprenoid compounds that exhibit antitumour activity. These compounds include tocotrienols and terpenoids (*e.g.*, limonene, geranoil, menthol and carvone). Overall, these compounds increase the tumour latency and decrease tumour multiplicity (Pandey, 2007 and 2014; Pandey and Madhuri, 2010e).

Chapter 6
Anticancer Plant-Drugs: Part III

Antibiotics for Chemotherapy of Cancers

'*Antibiotics*' are the chemical substances produced by various species of microorganisms (bacteria, fungi, actinomycetes) that suppress the growth of other microorganisms, and may eventually destroy them. The concept that substances derived from one living organism may kill another (antibiosis) is almost as old as the science of microbiology. The Chinese were aware, over 2500 years ago, of the therapeutic properties of moldy curd of soybean applied to carbuncles, boils and similar infections. The first investigators to recognize the therapeutic properties of microorganisms were Pasteur and Joubert, who recorded their observations and speculations in 1877. They noted that anthrax bacilli (*Bacillus anthracis* bacteria) grew rapidly when inoculated into sterile urine but failed to multiply and soon died if one of the '*common*' bacteria of the air was introduced in the urine at the same time. The '*golden age*' of antimicrobial therapy began with the production of penicillin in 1941, when this antibiotic was mass-produced and first made available for limited clinical trial. In 1928, while studying *Staphylococcus* (bacteria) variants in the laboratory at St. Mary's Hospital in London, Alexander Flemming observed that a mold contaminating one of the cultures caused the bacteria in its vicinity to undergo lysis. Broth in which the fungus grew was markedly inhibitory for many microorganisms. Because the mold belonged to the genus *Penicillium*, Flemming named this antibiotic as '*penicillin*'. Approximately 30 per cent of all hospitalized patients now receive one or more courses of antibiotic therapy and millions of potentially fatal infections are cured (Gilman *et al.*, 1985; Pandey and Madhuri, 2009b).

Certain Anticancer Antibiotics

Some microbial antibiotics (chemotherapeutic agents) isolated from different species of fungi for the treatment of various cancers include actinomycin D (dactinomycin), bleomycin, daunomycin, doxorubicin, epirubicin, idarubicin, geldanamycin and mitomycin C. Several data indicate that these antibiotics have been found very effective in clinical as well as experimental cases of many cancers/ tumours, *e.g.*, sarcoma, lymphoma, carcinoma, leukemia and others as discussed below (Chabner *et al.*, 2006; Gilman *et al.*, 1985; Pandey and Madhuri, 2009b; Rang *et al.*, 2003).

Actinomycin

The first antibiotic agent, isolated from a species of *Streptomyces* was actinomycin A. Many related antibiotics, including actinomycin D, have subsequently been obtained. Actinomycin D, marketed under the trade name dactinomycin, is one of the older chemotherapy drugs. It can be used for the treatment of various cancers, including neoplasms of childhood, choriocarcinoma, Kaposi's sarcoma, Ewing's tumour, testis tumour, soft tissues sarcomas, gestational trophoblastic neoplasia, Wilm's tumour and rhabdomyosarcoma. Actinomycin is a useful tool to determine the apoptosis, and distinguish dead cells and live cells. Actinomycin has been reported to inhibit the rapidly proliferating cells of normal and neoplastic origin. It binds DNA at the transcription initiation complex and prevents elongation by RNA polymerase. Thus, it blocks DNA dependent RNA synthesis. It can also interfere with DNA replication. Actinomycin D was the first antibiotic shown to have anticancer activity, but now it is not normally used as such, since it is highly toxic, causing damage to genetic material. It is much less effective by oral route than parenteral route. Its daily dose has been recommended to be 10 to 15 µg per kg, body weight, intravenously for 5 days.

Daunorubicin

This anthracycline antibiotic is among the most important antitumour agents. It is produced from the fungus *Streptomyces peuceticus*. Daunorubicin (daunomycin) is prescribed against some cancers, and is commonly used in specific types of leukemia or blood cancer (*e.g.*, acute myeloid and acute lymphocytic leukemias) and neuroblastoma. It has been reported to slow or stop the growth of cancer cells in the body as it intercalates with DNA. Many functions of DNA are affected, including blockade of DNA dependent RNA synthesis. The anticancer effect of daunorubicin can be achieved at the dose of 30 to 60 mg per square meter, intravenously, daily for 3 days or once weekly.

Doxorubicin

It is another important antitumour anthracycline antibiotic. Similar to daunorubicin, it is also obtained from the fungus *Streptomyces peuceticus*. Doxorubicin has been reported to be effective against soft tissue, oesteogenic and other sarcomas; Hodgkins's and non-Hodgkins's lymphomas; acute leukemia; cancers of breast, genitourinary, thyroid, lung and stomach; and neuroblastoma and other childhood sarcomas. It binds with DNA, thereby inhibits both DNA and RNA synthesis. Doxorubicin has been recommended in doses of 60 to 75 mg per square meter as a single intravenous infusion, and should be repeated after 21 days.

Epirubicin

Similar to daunorubicin and doxorubicin, epirubicin is also isolated from the fungus *Streptomyces peuceticus*. It is effective in breast cancer. Like doxorubicin, it also binds with DNA, leading to inhibition of both DNA and RNA synthesis.

Geldanamycin

This anticancer antibiotic is isolated from the fungus *Streptomyces hygroscopicus*. Geldanamycin has been reported to cause cell cycle disruption *and* inhibit the experimental cancer. Further, geldanamycin is a benzoquinone ansamycin antibiotic that binds to heat shock protein 90 (HSP90) and alters its function. HSP90 client proteins play important roles in the regulation of cell cycle, cell growth, cell survival, apoptosis, angiogenesis and oncogenesis. Thus, geldanamycin induces degradation of proteins that are mutated in tumour cells. This effect is mediated via HSP90.

Idarubicin (4-demethoxydaunorubicin)

It is an anthracycline antileukemic (anticancer) drug, isolated from the *Streptomyces peuceticus*. Similar to doxorubicin and epirubicin, idarubicin also binds with DNA, and inhibits both DNA and RNA synthesis. Clinically, it is given in cases of acute myeloid leukemia and breast cancer.

Bleomycin

It is produced from *Streptomyces peuceticus* (or *S. verticillus*) fungus. They have been found effective against various tumours, including squamous carcinoma of skin, cervix, head and neck, in addition to lymphomas (non-Hodgkin's lymphoma and Hodgkin's lymphoma) and testicular tumours. In comparison with many other anticancer agents, the bleomycins have minimal myelosuppressive and immunosuppressive activities. Bleomycin has been reported to degrade the performed DNA. In fact, it causes the accumulation of cells in the G_2 phase of cell cycle and many of these cells display chromosomal aberrations, including chromatid breaks, gaps and fragments as well as translocations. It also causes *'free radicals'* to be made inside the body. Free radicals are hyperactive oxygen atoms that damage DNA. Bleomycin is usually administered parenterally and its recommended dosage is 10 to 20 units per square meter, intramuscular (or intravenous, subcutaneous or intrathecal injection), weekly or twice weekly.

Mitomycin

The mitomycins are a family of aziridine-containing natural products, isolated from *Streptomyces lavendulae* (or *S. caespitosus*). One of these compounds, mitomycin C, finds use as a chemotherapeutic agent by virtue of its antitumour antibiotic activity. Mitomycin C has been employed to treat upper gastrointestinal (*e.g.*, oesophageal), breast, gastric, colorectal, bladder, anal and lung cancers. Total synthesis of mitomycins via isomitomycin was explained by Fukuyama and Yang, and mitomycin C was stated to be a small, fast and deadly (but very selective) antibiotic; however, the molecular characterization and analysis of the biosynthetic gene cluster for mitomycin C from *S. lavendulae* was done by Mao and co-workers. Mitomycin C has been reported to be a potent DNA cross linker. A single crosslink per genome has been effective in killing bacteria. This is accomplished by reductive activation of enzyme, followed by two N-alkylations. Mitomycin C is recommended in dosage of 2 mg per square meter, intravenously, daily for 5 days. This course should be repeated after a 2-day interval.

Phytoestrogen Against Cancer

'Phytoestrogens' (also known as *'phytosterols'*) are a group of phytochemicals or natural plant compounds that can act like estrogen hormone. Estrogen is a hormone necessary for childbearing, and is involved with bone and heart health in women. *'Phyto'* is from Greek origin that is generally used as a prefix to describe substances derived from plants. Phytoestrogens, therefore, are substances from plants which have estrogen-like qualities (Pandey and Madhuri, 2011b; Pandey *et al.*, 2013a).

The prevailing scientific opinion on the phytoestrogens is favourable as there is substantial research indicating that they hold great potential for health benefits. The epidemiological studies have suggested that typical Asian diets, which have always been much higher in these substances than Western diets, appear to be associated with a significantly lower risk of breast, prostate and colon cancer as well as a reduced incidence of heart disease and osteoporosis. These degenerative diseases have in fact long been associated with modern Western diets and, therefore, the studies have truly caused no surprise by reinforcing this idea and favouring Asian diets for good health. The benefits of phytoestrogens to good health are mainly due to the effects they have on the body's hormonal balance, acting as both agonists and antagonists. To understand how these substances help the body's hormonal balance, it is useful to recall what hormones are and how they work. Hormones are chemical substances produced by the body's endocrine glands and released into the bloodstream to act as chemical messengers, travelling through the body with instructions to trigger activity in their target tissues. These target tissues each contain receptor sites specific to a particular hormone and the required effect is initiated when the hormone in question arrives and docks at those receptor sites. For example, estrogens are released from the ovaries and travel through the blood to the breast area where they dock and deliver the instruction to initiate mammogenesis (Pandey and Madhuri, 2011b).

About 50 years ago, researchers became aware that phytoestrogens present in alfalfa and clovers could affect the fertility of livestock. Now, multiple epidemiological studies showed a relationship between high dietary intake of isoflavones and lignans, and lower rates of certain cancers, cardiovascular problems and menopausal symptoms. As far back as 1985, it was known that phytoestrogens could compete with estradiol for binding to intercellular ERs. Although, scientific evidence suggests that phytoestrogens may have a role in preventing chronic diseases. A strong body of evidence suggests that they may be effective in preventing and treating prostate cancer, due to their antiandrogenic properties (Pandey and Madhuri, 2011b; Pandey *et al.*, 2013a).

Plants for Foods with Phytoestrogens

Phytoestrogens have been found in more than 300 types of foods. Most of the food phytoestrogens are from one of two chemical classes: (a) Flavonoids (*e.g.*, isoflavonoids or isoflavones, flavones and coumestans or coumestrols); (b) lignans (*e.g.*, secoisolariciresinol, matairesinol, pinoresinol and lariciresinol). Isoflavonoids or isoflavones (genistein, daidzein, glycitein and formononetin) are present in highest amounts in soybean, soybean products (*e.g.*, tofu), alfalfa, linseed, fenugreek, fennel and red clover; soy being a major potential source of human exposure. Lignans are found in higher fibre foods such as cereal brans and beans; flax seeds contain large amounts of lignans. Coumestans (coumestrols) are present in various beans such as split peas, pinto beans and lima beans; alfalfa and clover sprout foods contain highest amount of coumestans. Estrogen is available in medically formulated pills. However, dietary estrogen (phytoestrogen) can be also present in a variety of food products (including herbs). The Canadian researchers analyzed 121 food samples, of which the food samples with the highest total phytoestrogen content were nuts and oil seeds, followed by soy products. The two most important soy isoflavones at present are genistein and daidzein. But lignans are also a very important source of these subsances in the British diet as they are present in most fibre providing foods (Pandey and Madhuri, 2011b).

Usefulness of Phytoestrogens in Cancers and other Diseases

The studies have indicated that phytoestrogens may provide good health, including potential reduction in the cancers of breast, prostate and colon, and cardiovascular disease risks; and possible protection against osteoporosis (bone loss) and menopausal symptoms. Besides, both flavonoid and lignan phytoestrogens have antioxidant activity (Pandey and Madhuri, 2011b). Some bone-sparing effects of phytoestrogens have been demonstrated with natural and commercial phytoestrogens and the synthetic ipriflavone through limited studies. Research has yielded positive, yet inconsistent, trends with respect to bone turnover markers, bone mineral density and bone mineral content, but little fracture data exists. Adequate phytoestrogen dosages for osteoporosis have yet to be determined; except 200 mg of ipriflavone three times a day is an established dosage in most trials. Inconsistency in positive results also exists for menopausal symptoms and breast cancer, although evidence is established for reduction of cardiovascular disease (Button and Patel, 2004).

The effect of phytoestrogens against cancer has been seen in animal studies, especially when exposed during breast development. Isoflavonoids and lignans stimulate the proliferation of ER and breast cancer cells. Hence, the phytoestrogens at high concentrations inhibit cell growth. The antiangiogenic (anticardiac) effects of genistein, daidzein and biochanin A may contribute to antitumor activity. Consumption of soy isoflavones at the rate of 30 mg per day may reduce hot flashes by 30 to 50 per cent. Isoflavone intake increases bone mineral density and can be useful in preventing post-menopausal osteoporosis. Average intake of 47 g per day soy protein results in 9 per cent decrease in total cholesterol, 13 per cent decrease in LDL cholesterol and a trend towards HDL cholesterol. Flavanoids decrease the platelet aggregation; the genistein-induced inhibition of growth factor activity can interfere with platelet and thrombin action. Phytoestrogens also regulate the delayed menstrual cycle in women as they reduce LH, FSH and progesterone. The action of phytoestrogen genistein on the uterus and vagina of ovariectomized DA/Han rats after 3 day oral administration (25, 50 or 100 mg/kg/day) was compared to EO (0.1

mg/kg/day). A dose dependent increase of uterine wet weight and uterine and vaginal epithelial height, a dose dependent up-regulation of complement C_3, down-regulation of clusterin mRNA expression and a stimulation of vaginal cornification was observed. Uterine gene expression and vaginal epithelium respond to genistein at doses where no significant effects on uterine wet weight were detectable. In general, the vagina was more sensitive to genistein than the uterus. To analyse the action of genistein in malignant uterine tissue, the impact of a 28 day treatment with 50 mg per kg per day of genistein on the *in vivo* tumour growth of endometrial adenocarcinoma cells, following subcutaneous inoculation into syngeneic DA/Han rats, was assessed. In contrast to EO (0.1 mg/kg/day), a dose of 50 mg per kg per day of genistein did not affect tumour growth. In conclusion, four independent uterine and vaginal parameters indicated that genistein is a weak ER agonist in the uterus and vagina of female DA/Han rats, and evidence provided for a selective ER modulator-like action of genistein in normal and malignant uterine tissue (Pandey *et al.*, 2013a).

Based on results from some *in vitro* and animal studies, concern has arisen that the estrogen agonist effects of isoflavones might increase the growth of breast cancer cells. Though there is still some controversy, the majority of scientific opinions indicate that phytoestrogen-containing foods prevent and treat the breast cancer. Several studies have indicated that countries with the highest phytoestrogen consumption have the lowest rates of breast cancer. No studies have found an increased risk of breast cancer with increased soy consumption. Many *in vitro* experiments detected anticancer effects from phytoestrogens at high concentrations (but mild stimulatory effects at lower concentrations). Several reports have indicated that exposure of young rats (but not adult rats) to genistein results in a large reduction in mammary cancer later in life. One human study found a similar protective pattern for women who ate tofu as teenagers. Several studies quoted that phytoestrogens have antiangiogenesis effects, discouraging the growth of new blood vessels that tumours need for survival. Equol is a metabolite of diadzein, glucoside form of daidzein. It is produced by some 30 to 40 per cent of people who ingest the isoflavone. In ER assays, equol exhibits roughly the same binding affinity as genistein; however, it tends to stay in circulation longer, presumably increasing exposure of tissues to its effects. The ability to produce equol seems to be genetic and not influenced by diet. Reports suggest that people who produce equol have hormonal profiles associated with a lower risk of breast cancer: lower concentrations of androstenedione, dehydroepiandrosterone, estrone, cortisol, and testosterone; and higher concentrations of sex hormone binding globulin. Lignans have demonstrated beneficial effects with breast, prostate, and colon cancer as well as with hypercholesterolemic atherosclerosis and chronic kidney disease. In the colon, bacteria convert the botanical lignans into the mammalian lignans enterodiol and enterolactone. Evidence suggests that a healthy colon flora population may be necessary for humans to derive significant benefit from lignans. *In vitro*, lignans have been demonstrated to bind to sex hormone binding globulin, displacing estradiol and testosterone. Several animal studies showed that lignans have significant anticarcinogenic effects. The latest research indicates that high levels of lignans are associated with lower breast cancer risk (Pandey and Madhuri, 2011b).

Numerous epidemiological studies have shown an inverse correlation between cancer incidence and fruit and vegetable consumption (Pandey and Madhuri, 2010e); lignans are among the many compounds likely to be responsible for this effect. It has also been demonstrated that women with breast cancer have lower plasma levels of lignans than women without breast cancer. Further, another study compared the neurotrophic effects of six different isoflavones to the effect of estradiol in order to determine if the isoflavones had estrogen agonist properties in cultured human hippocampal cells. Estradiol protected neuronal mitochondria from damage and promoted neuron process outgrowth (a cellular correlate of memory). The phytoestrogens, however, demonstrated a modest protective effect on the cell membranes, which the researchers suspected was due to their antioxidant properties.

Numerous studies have stated that isoflavones can affect the brain metabolism and neurological performance of mice and rats. Furthermore, soy has long been known to have effects on the thyroid. Isoflavones in soy (and flavonoids from other sources as well) inhibit the enzyme thyroid peroxidase, which is involved in thyroid hormone synthesis. This study explored the inhibitory effects of genistein and daidzein, which were completely reversed with the addition of sufficient iodine. Clinical problems from ingesting high levels of phytoestrogens, such as aggravated hypothyroidism or goiter, can occur in iodine-deficient or hypothyroid individuals. There is some speculation that soy formula could be contributing to the increase in premature puberty among American girls, but scientific data is lacking. Phytoestrogens exhibit the antioxidant activity *in vitro* and *in vivo*, hence they act against cancer (Pandey *et al.*, 2013a).

Mechanism of Action of Phytoestrogens

Phytoestrogens may act by many ways in the body. The chemical structure of phytoestrogens is similar to estrogen, and they may act as mimics (copies) of estrogen. On the other hand, phytoestrogens also have effects that are different from those of estrogen. Working as estrogen mimics, phytoestrogens may either have the same effects as estrogen or block estrogen's effects. Which effect the phytoestrogen produces can depend on its dose. The phytoestrogen can act like estrogen at low doses but block estrogen at high doses. Estrogen activates a family of proteins called 'ERs'. The studies have shown that phytoestrogens interact more with some members of the ER family. Phytoestrogens acting differently from estrogen, may affect communication pathways between cells, prevent the formation of blood vessels to tumours or alter processes involved in the processing of DNA for cell multiplication. Which of these effects occur is unknown. In the breast, phytoestrogens travel to deliver instruction to initiate the mammogenesis and for that reason they are used in herbal preparations for natural breast enhancement or natural breast enlargement (Pandey *et al.*, 2013a).

Phytoestrogens are thought to act as estrogen agonists by occupying ER sites when natural estrogens are unavailable. For example, the body's natural estrogen levels inevitably decline with the onset of menopause and phytoestrogens may help to offset this decline if they can fill receptor sites instead. Once docked on the sites, they exert estrogen-like activity and may initiate the required effect just as natural estrogen would have done. Phytoestrogens are also thought to act as estrogen antagonists by occupying ER sites ahead of the body's natural estrogens and equally importantly ahead of synthetic estrogens, and also environmental estrogens derived from chemical products, otherwise known as *'bad estrogens'* or *'xenoestrogens'*. In order words, where estrogen levels are high, phytoestrogens are able to compete with the bod's natural estrogens or the artificial estrogens present and may fill the receptor sites before they do. If this happens, they will decrease the estrogenic activity in the body, because the effect of docked phytochemicals on the target tissues will be less than if the available estrogens had been allowed to dock. A limitation on the hormone balancing actions of phytoestrogens is that they do not have estrogenic properties when still in the plant or even at the time they are consumed, but acquire them only during the digestive process through the actions of bacteria residing in the GIT. The bacteria cause phytochemicals to undergo complex metabolic conversions, leading to the formation of an estrogen-like metabolite which can then be absorbed by the body. This means that the biologically useful forms of phytoestrogens, the metabolites, are dependent for their existence upon a digestive system in good order and complete with adequate microflora capable of converting the basic phytochemical into active forms. This is a limitation on the effectiveness of phytoestrogens for simple reason that there are a number of factors that can adversely affect the stability of normal gastrointestinal flora. Poor or inappropriate diets, stress and antibiotics can all significantly disrupt the ideal healthy balance of gastrointestinal organisms. Antibiotics can quickly destroy friendly bacteria, as well as invading organisms they are actually meant to destroy. High fat intake is another culprit. However, a high-fibre diet can help the metabolism of phytochemicals (Pandey and Madhuri, 2011b; Pandey *et al.*, 2013a).

Recent research suggests that phytoestrogens may be natural *'selective estrogen receptor modulators'*, which means that they can bind to certain ERs in some tissues, either activating or down-regulating cellular responses. Depending on concentrations of endogenous estrogens, as well as on which receptor complexes are activated or down-regulated, the selective estrogen receptor modulators can have either estrogenic or antiestrogenic effects. Through preferential binding to β-ER, phytoestrogens activate cardioprotective and bone-stabilizing metabolic processes. Simultaneously, they appear to down-regulate the activity of α-ER prominent in breast and uterine tissue. This is one possible mechanism behind their proposed anticancer effects. In addition, phytoestrogens can favourably affect the balance of estrogen metabolites in the body. *'Bad'* metabolites (16 α-hydroxyestrone, 4-hydroxyestrone and 4-hydroxyestradiol) are genotoxic and mutagenic. The ratio of *'good'* (2-hydroxyestrone) to *'bad'* metabolites is increasingly being used as a marker to assess cancer risk. Non-ER mediated effects on growth regulation in human breast cancer cells have also been documented for genistein. Many *in vitro* studies have indicated that phytoestrogens have some 1/100 to 1/1000 the binding affinity of estradiol for cellular ERs. This has led to the interpretation that phytoestrogens are 100 to 1000 times weaker than estradiol. In addition to binding affinity, another factor to consider is the influence of high plasma levels of phytoestrogens which can be present at some 100 to 1,000 times the concentration of endogenous estrogens (and even higher in soy-formula-fed infants). In addition, phytoestrogens may have different bioavailabilities than endogenous estrogens, due to the fact that they bind less tightly to steroid hormone serum transport proteins. This is because that many phytoestrogens are converted by human colon bacteria into other compounds (including enterodiol, enterolactone and equol). Some of these metabolites are more potent than their precursors, while others are less so. Different individuals, depending on factors such as their particular gut flora and/or genetic makeup, produce different concentrations and proportions of these metabolites. Also, phytoestrogen activity is modulated by the levels of a person's endogenous estrogens. Further, the estrogenic effect of any particular compound is not the same in different types of cells and tissues; nor is it identical in different species, so it is not possible to directly apply the results of *in vitro* and animal research to humans. Finally, the different sexes (in both animals and humans) can have different responses to phytoestrogens. Receptor-binding affinity, then, is only one factor amongst many that determines actual hormonal effects of any particular phytochemical. Overall, the situation is far more complex; many biochemical factors are involved; all phytoestrogens are not the same; all tissues do not respond identically; some people respond differently than others. There is also evidence that isoflavones and lignans exert anticancer effects through other mechanisms, independent of their interactions with ER. For example, isoflavones at physiological concentrations have been found to inhibit an enzyme which catalyzes the transformation of weaker estrogen, estrone, into more cancer-promoting estradiol. Another study found that phytoestrogens inhibit a second enzyme important in steroid biosynthesis. Isoflavones also exhibit some antioxidant activity, which may contribute to anticancer effect (Pandey and Madhuri, 2011b).

Ocimum sanctum against Cancer

Importance and Various Uses of *Ocimum sanctum*

Known as 'Tulsi' in Hindi and as 'Holy Basil' in English, the botanical plant *Ocimum sanctum* Linn. belongs to plant family Lamiaceae. It has made important contribution to the field of science from ancient times as also to modern research due to its large number of medicinal properties. Tulsi is an important symbol of the Hindu religious tradition. Although the word 'Tulsi' gives the connotation of the incomparable one, its other name, Vishnupriya means the one that pleases Lord Vishnu. Found in most of the Indian homes and worshipped, its legend has permeated Indian ethos down the ages. Tulsi has been described as of two types: '*Vanya*' (wild) and '*Gramya*' (grown in homes). Although having identical usage, the former has darker leaves. Tulsi is a popular home remedy for many ailments such as wound, bronchitis, liver diseases, catarrhal fever, otalgia, lumbago, hiccough, ophthalmia, gastric disorders, genitourinary disorders, skin diseases, various forms of poisoning and psychosomatic stress disorders (Das and Vasudevan, 2006; Pandey and Madhuri, 2010d; Prajapati *et al.*, 2003). *O. sanctum* also acts as aromatic, stomachic, carminative, demulcent, diaphoretic, diuretic, expectorant, alexiteric, vermifuge and febrifuge (Gupta *et al.*, 2002; Pandey and Madhuri, 2010d).

Traditionally, Tulsi is known as '*the elixir of life*' since it promotes longevity. Different parts of its plant are used in Ayurveda and Siddha Systems of Medicine for prevention and cure of many illnesses and everyday ailments like common cold, headache, cough, flu, earache, fever, colic pain, sore throat, bronchitis, asthma, hepatic diseases, malaria fever, as an antidote for snake bite and scorpion sting, flatulence, migraine headaches, fatigue, skin diseases, wound, insomnia, arthritis, digestive disorders, night blindness, diarrhea and influenza. The leaves are good for nerves and to sharpen memory. Chewing of Tulsi leaves also cures ulcers and infections of mouth (Pandey and Madhuri, 2010d; Prajapati *et al.*, 2003).

O. sanctum has been adored in almost all ancient Ayurvedic texts for its extraordinary medicinal properties. It is pungent and bitter in taste and hot, light and dry in effect. Its seeds are considered to be cold in effect. The roots, leaves and seeds of Tulsi possess several medicinal properties. Ayurvedic texts categorise *O. sanctum* as stimulant, aromatic and antipyretic. While alleviating kapha and vata, it aggravates pitta. It has a wide range of action on the human body mainly as a cough alleviator, a sweat-inducer and a mitigator of indigestion and anorexia. *O. sanctum* has been reported to possess a variety of biological/pharmacological activities such as antibacterial, antiviral, antifungal, antiprotozoal, antimalarial, anthelmentic, antidiarrhoeal, analgesic, antipyretic, antiinflammatory, antiallergic, antihypertensive, cardioprotective, CNS depressant, memory enhancer, antihypercholesterolaemic, hepatoprotective, antidiabetic, antiasthmatic, antithyroidic, antioxidant, anticancer, chemopreventive, radioprotective, immunomodulatory, antifertility, antiulcer, antiarthritic, adaptogenic/antistress, anticataract, antileucodermal and anticoagulant activities (Pandey and Madhuri, 2010d).

Phytoconstituents of *Ocimum sanctum*

O. sanctum leaves contain 0.7 per cent volatile oils, comprising about 71 per cent eugenol and 20 per cent methyl eugenol. Oils also contain carvacrol and sesquiterpine hydrocarbon caryophyllene. Fresh leaf and stem of *O. sanctum* extract yield some phenolic compounds (antioxidants) such as cirsilineol, cirsimaritin, isothymusin, apigenin and rosamarinic acid, and appreciable quantities of eugenol. Two flavonoids, *viz.*, orientin and vicenin from AqE of *O. sanctum* leaves have been isolated. Ursolic acid, apigenin, luteolin, apigenin-7-O-glucuronide, luteolin-7-O-glucuronide, orientin and molludistin have also been isolated from the leaf extract. *O. sanctum* also contains a number of sesquiterpenes and monoterpenes, *viz.*, bornyl acetate, β-elemene, neral, α- and β-pinenes, camphene, campesterol, cholesterol, stigmasterol and β-sitosterol (Pandey and Madhuri, 2010d).

Anticancer and Related Activities of *Ocimum sanctum*

Anticancer Activity

The anticancer activity of *O. sanctum* has been cited in different reports (Madhuri, 2008; Madhuri and Pandey, 2010a; Pandey, 2009; Pandey and Madhuri, 2006a; 2006b and 2010d). The AlE of *O. sanctum* leaves has a modulatory influence on carcinogen metabolizing enzymes, *e.g.*, cytochrome P450, cytochrome b_5, aryl hydrocarbon hydroxylase and GST, which are important in detoxification of carcinogens and mutagens. The anticancer activity of *O. sanctum* has been reported against human fibrosarcoma cells culture, wherein AlE of this drug induced the cytotoxicity at the dose of 50 µg per ml and above. Morphologically, the cells showed shrunken cytoplasm and condensed nuclei. The DNA became fragmented on observation in agarose gel electrophorosis. *O. sanctum* significantly decreased the incidence of B(a)P induced neoplasia of fore-stomach of mice and 3'-methyl-4-dimethylaminoazo-benzene induced hepatomas in rats. The AlE of the leaves of *O. sanctum* was shown to have an inhibitory effect on chemically induced skin papillomas in mice. Topical treatment of Tulsi leaf extract in DMBA induced papillomagenesis significantly reduced the tumour incidence, average number of papillomas/mouse and cumulative number of papillomas in mice. Topical application of the extract significantly elevated reduced GSH content and GST activities. A similar activity was observed for eugenol, a flavonoid present in many plants, including Tulsi. Oral treatment of fresh leaves paste of Tulsi may have the ability to prevent the early events of DMBA induced buccal pouch carcinogenesis. Leaf extract of *O. sanctum* blocks or suppresses the events associated with chemical carcinogenesis by inhibiting metabolic activation of the carcinogen. The anticancer activity of *O. sanctum* was also noticed in Swiss albino mice bearing EAC and S180 tumours.

Antioxidant Activity

The antioxidant activity of *O. sanctum* has been reported by many authors (Madhuri, 2008; Madhuri and Pandey, 2010a; Pandey, 2009; Pandey and Madhuri, 2006a; 2006b and 2010d). The antioxidant properties of flavonoids and their relation to membrane protection have been found. Antioxidant activity of the flavonoids (orientin and vicenin) *in vivo* was expressed in a significant reduction in the radiation induced lipid peroxidation in mouse liver. *O. sanctum* extract has significant ability to scavenge highly reactive free radicals. The phenolic compounds, *viz.*, cirsilineol, cirsimaritin, isothymusin, apigenin and rosmarinic acid, and appreciable quantities of eugenol (a major component of the volatile oil) from *O. sanctum* extract of fresh leaves and stems possessed good antioxidant activity.

Chemopreventive Activity

The chemopreventive effect of *O. sanctum* leaf extract is probably through the induction of hepatic/extrahepatic GST in mice. Elevated levels of reduced GSH in liver, lung and stomach tissues in *O. sanctum* extract supplemented mice were also found. Significant antiproliferative and chemopreventive activities were observed in mice with high concentration of *O. sanctum* seed oil. The potential chemopreventive activity of seed oil has been partly attributed to its antioxidant activity (Pandey and Madhuri, 2010d).

Radioprotective Activity

The radioprotective effect of *O. sanctum* was significantly reported by Uma Devi and coworker in the year 1995. Two isolated flavonoids, *viz.*, orientin and vicenin from *O. sanctum* leaves showed better radioprotective effect as compared with synthetic radioprotectors. These have shown significant protection to the human lymphocytes against the clastogenic effect of radiation at low, non-toxic concentrations. The combination of *O. sanctum* leaf extract with WR-2721 (a synthetic radioprotector) resulting in higher bone marrow cell protection and reduction in the toxicity of WR-

2721 at higher doses, suggested that the combination would have promising radioprotection in humans (Pandey and Madhuri, 2010d).

Immunomodulatory Activity

Steam distilled extract from the fresh leaves of *O. sanctum* showed modification in the humoral immune response in albino rats which could be attributed to such mechanisms as antibody production, release of mediators of hypersensitivity reactions and tissues responses to these mediators in the target organs (Pandey and Madhuri, 2010d). *O. sanctum* seed oil appears to modulate both humoral and cell-mediated immune responsiveness and GABAergic pathways may mediate these immunomodulatory effects (Mukherjee *et al.*, 2005).

Withania somnifera as an Anticancer Drug

Various Uses of *Withania somnifera*

Withania somnifera (Ashwagandha, Asgandh, Indian ginseng, winter cherry) belongs to Solanaceae plant family. This has been an important herb in the Ayurvedic and indigenous medicine systems for over 3000 years (Madhuri and Pandey, 2009c). Various parts of herb are traditionally used in different diseases. Its root is traditionally used as aphrodisiac, liver tonic, antiinflammatory, astringent and antidepressant and in impaired memory, neurasthenic, poor muscle tone and fever (Jha, 2007). Western research supports its polypharmaceutical use, confirming antioxidant, antiinflammatory, immunomodulating and antistress properties in the whole plant extract and several separate constituents (Mishra *et al.*, 2000). Its root is a potential source of hypoglycaemic, diuretic and hypocholesterolemic drugs (Andallu and Radhika, 2000). Methanol and hexane extracts of the leaves and roots of *W. somnifera* showed potent antibacterial activity (Arora *et al.*, 2004). The leaf, root and root bark of *W. somnifera* showed antimalarial activity (Dikasso *et al.*, 2007).

Anxiolytic and antidepressant actions of the glycowithanolides, isolated from the roots of *W. somnifera*, in rats were assessed. *W. somnifera* inhibited stress-induced gastric ulcer more effectively as compared to the standard drug ranitidine. It significantly inhibited haloperidol or reserpine-induced catalepsy and provided hope for the treatment of Parkinson's disease. It also possesses antivenom and antiinflammatory activities. It also exhibits antiangiogenic effect. Experimentally, its cardioprotective effect has been reported. In acute and chronic seizure models of animals, *W. somnifera* has acted as an anticonvulsant drug. The glycoprotein of *W. somnifera* has antifungal activity. It also possesses adaptogenic, cardiotropic and anticoagulant properties. Withaferin A, a major chemical constituent of *W. somnifera*, has antiarthritic, antimicrobial, antimitotic and viricide activity. Besides these applications, *W. somnifera* has also been reported to possess anticancer activity (Madhuri and Pandey, 2009c; Pandey *et al.*, 2013a).

Phytoconstituents of *Withania somnifera*

The majority of the phytoconstituents of *W. somnifera* are withanolides (steroidal lactones with ergostane skeleton) and alkaloids. These include withanone, withaferin A, and several other withanolides and withasonidienone (IDMA, 2002; Pandey *et al.*,

2013a). Much of the pharmacological activity of *W. somnifera* has been attributed to two main withanolides, withaferin A and withanolide D (Gupta and Rana, 2007). Apart from these contents, this plant also contains chemical constituents like withaniol, acylsteryl glucosides, starch, reducing sugar, hentriacotane, ducitol, a variety of amino acids including aspartic acid, proline, tyrosine, alanine, glycine, glutamic acid, cystine, tryptophan and high amount of iron (Jha, 2007; Pandey *et al.*, 2013a; Prajapati *et al.*, 2003). Most of these compounds have been found in both aqueous and alcoholic (ethanol) extracts of *W. somnifera* root. The phytochemicals like alkaloids, reducing sugars, resins, saponins, fixed oils, anthraquinones, proteins and amino acids have been found in the AqE and AlE, while glycosides have been determined in the AlE of *W. somnifera* root (Madhuri and Pandey, 2009c; Somkuwar, 2003).

Anticancer and Related Activities of *Withania somnifera*

The AlE of the whole plant of *W. somnifera* (200 mg/kg, orally, daily for 7 months) reduced the tumour incidence significantly against urethane (125 mg/kg, biweekly for 7 months) induced lung adenomas in adult male albino mice. The histological appearance of lungs protected by *W. somnifera* was similar to those of the lungs of normal mice. This drug has been found to scavenge the reactive molecules, leading to antimutagenesis and anticarcinogenesis. The AlE of the root of *W. somnifera* at daily doses of 200 to 1000 mg per kg, ip for 15 days starting from 24 hr after intradermal inoculation of 5×10(5) cells of S180 in Balb/c mice. The cumulative doses of 7.5 to 10 g, ip at daily doses of 500 or 750 mg per kg seemed to produce a good response in this tumour. The antitumour effect of root extract of *W. somnifera* and its modification by heat were studied *in vivo* on S180 tumour grown on the dorsum of adult Balb/c mouse. The AlE of the root of this plant produced *in vivo* growth inhibitory effect on transplantable mouse tumour, S180. Further, the AlE of the dried root of this plant as well as its active component withaferin A showed significant antitumour effect in experimental tumours *in vivo*. The anticancer effect of withaferin A was also noted against EAC *in vivo*. In another study, withaferin A inhibited the tumour growth and increased tumour free survival in a dose-depended manner (Madhuri and Pandey, 2009c; Uma Devi, 1996; Uma Devi *et al.*, 2000).

W. somnifera may be used as an adjuvant during cancer chemotherapy for prevention of bone marrow depression associated with anticancer drugs (Gupta *et al.*, 2001). Administration of the root extract (20 mg/dose/animal, ip) of *W. somnifera* was found to inhibit the 20-methylcholanthrene induced sarcoma in mice, and increase the life span of tumour bearing animals. Administration of this extract also reduced the skin carcinogenesis induced by DMBA and croton oil. *W. somnifera* treated animals showed increased GSH, SOD, glutathione peroxidase and catalase in the liver and skin. Methanol extracts of *W. somnifera* root at a dose of 65 µg per ml or 265 µg per ml were able to down-regulate the expression of p34cdc2, a cell-cycle regulatory protein. This protein is expressed during cellular proliferation, and down regulation arrests the cell cycle in the G_2/M transition phase. This plant showed a significant increase in cytotoxic T lymphocyte production both *in vivo* as well as *in vitro*, and it may reduce tumour growth. *W. somnifera* (root extract) treated splenocytes along with the mitogen lipopolysaccharide could stimulate the lymphocyte proliferation six times more than the normal. The NK cell activity was enhanced significantly in both the normal and tumour bearing animals (Davis and Kuttan, 2002; Madhuri and Pandey, 2009c). Oral treatment with hydroalcoholic extract (HAE) of *W. somnifera* root at the dose of 400 mg per kg (one week before injecting 20-methylcholanthrene and continued until 15 weeks thereafter) significantly reduced the tumour incidence, tumour volume and enhanced the survival of the mice, compared with 20-methylcholanthrene injected mice bearing with fibrosarcoma. The chemopreventive effect was demonstrated in a study of HAE of *W. somnifera* root on DMBA induced skin cancer in mice. A significant decrease in incidence and average number of skin lesions was noticed. Additionally, levels of reduced GSH, SOD, catalase and glutathione peroxidase in exposed tissue returned to near normal values, following the administration of extract. The chemopreventive activity is thought to be due in part to antioxidant/free radical scavenging activity of the extract (Madhuri and Pandey, 2009c; Prakash *et al.*, 2002).

W. *somnifera* root showed anticancer effect activity in Swiss albino mice bearing EAC and S180 tumours. Withaferin A from *W. somnifera* was most effective at delaying tumour growth and doubling time in fibrosarcoma (Madhuri and Pandey, 2009c; Somkuwar, 2003; Uma Devi *et al.*, 2000). *In vitro*, withanolides inhibited the growth in human breast (MCF-7), CNS (SF-268), lung (NCL-H460) and colon (HCT-116) cancer cell lines comparable to doxorubicin. Withaferin A more effectively inhibited the growth of breast and colon cancer cell lines than did doxorubicin. These results suggest that the extracts of *W. somnifera* root may prevent or inhibit tumour growth in cancer patients, and suggest a potential for development of new chemotherapeutic agents (Jayaprakasam *et al.*, 2003). The AlE of *W. somnifera* is effective against different prostate cancer cell lines of various metastatic potential (Rao *et al.*, 2004). Both *in vivo* and *in vitro* research attest to cytotoxic and antitumour potential of *W. somnifera*. The osteogenic sarcoma and breast carcinoma cell lines were treated with AqE of *W. somnifera* leaf powder in doses of 3 to 24 µg per ml. These cancers exposed to high oxidative stress via a high-glucose medium or exposure to hydrogen peroxide were more susceptible to oxidative damage after treatment with *W. somnifera* extract, suggesting that drug has antiproliferative effect on tumour cells (Kaur *et al.*, 2004; Pandey *et al.*, 2013a).

A significant increase in the life span and a decrease in the cancer cell number and tumour weight were noted in the tumour-induced mice after treatment with AlE of *W. somnifera* root. The haematological parameters were also corrected. These observations are suggestive of the protective effect of *W. somnifera* in DAL (Christina *et al.*, 2004). Pretreatment to animals with 1-oxo-5β,6 β-epoxy-witha-2-enolide (20 mg/kg), isolated from the roots of *W. somnifera*, prior to exposing the animals to UV radiation, prevented the incidence of skin carcinoma. Pretreatment of *W. somnifera* root powder extract (20 mg/dose/animal/24 hr, ip for 10 days) or a constituent of *W. somnifera*, withanolide D (500 µg/dose/animal/24 hr, ip for 10 days) resulted in a significant reduction in tumour (melanoma) and increase in life span of mice. *W. somnifera* inhibited B(a)P induced fore-stomuch papillomagenesis, showing up to 60 per cent and 92 per cent inhibition in tumour incidence and multiplicity, respectively. Similarly, *W. somnifera* inhibited the DMBA induced skin papillomagenesis, showing up to 45 per cent and 71 per cent inhibition in tumour incidence and multiplicity. The B(a)P induced cancer animals were treated with *W. somnifera* extract for 30 days, resulting into significant alteration of the levels of immunocompetent cells, immune complexes and immunoglobulins. It was also observed that the combination chemotherapy of *W. somnifera* along with paclitaxel is a promising chemotherapeutic agent against lung cancer induced by B(a)P in swiss albino mice. The antiproliferative activity was screened against human laryngeal carcinoma (Hep2) cells by microculture tetrazolium assay. The effect was confirmed *in vivo* by mouse sponge implantation method. The experiments suggest that the roots of *W. somnifera* possess cell cycle disruption and antiangiogenic activity, which may be a mediator for its anticancer action (Mathur *et al.*, 2006; Pandey *et al.*, 2013a).

Withanolides inhibited the cyclooxygenase enzymes, lipid peroxidation and proliferation of tumour cells. It suppressed NF-kappaB activation induced by a variety of inflammatory and carcinogenic agents including tumour necrosis factor, interleukin-1β, doxorubicin and cigarette smoke condensate. Suppression was not cell type specific, as both inducible and constitutive NF-kappaB activation were blocked by withanolides. The suppression occurred through the inhibition of inhibitory subunit of I-kappaB α-kinase activation, I-kappaB α-phosphorylation, I-kappaB α-degradation, p65 phosphorylation and subsequent p65 nuclear translocation. Overall, withanolides inhibit the activation of NF-kappaB and NF-kappaB-regulated gene expression, which may explain the ability of withanolides to enhance apoptosis and inhibit invasion and osteoclastogenesis (Ichikawa *et al.*, 2006). The AqE of *W. somnifera* had inhibitory effect on Chinese hamster ovary cell lines (Sumanthran *et al.*, 2007). Withaferin A induced the prostate apoptosis responses-4 in androgen-refractory prostate cancer cells, showing that *W. somnifera* exhibits the cytotoxic effect against variety of cancer cell lines (Srinivasan *et al.*, 2007). *W. somnifera* primarily induced the oxidative stress in human leukemia HL60 cell and in several other cancer cell lines (Malik *et al.*, 2007). Further, withaferin A inhibited DNA binding of NF-kappaB and caused nuclear cleavage of P65/Rel by activated caspase-3. N acetyl-cysteine rescued all these events suggesting thereby a pro-oxidant effect of withaferin A. The anticancer effect of *W. somnifera* has also been cited by other workers (Gupta and Rana, 2007; Kaur *et al.*, 2004; Madhuri and Pandey, 2007 and 2008; Madhuri and Pandey, 2009c; Pandey and Madhuri, 2006a and 2006b; Pandey *et al.*, 2013a).

Possible Anticancer Mechanism of *Withania somnifera*

Anticancer activity of *W. somnifera* is related to its multiple functions. *W. somnifera* may increase the overall effectiveness of cancer treatment. Its anticancer activity is probably due to action of its main constituents, *viz.*, withaferin A (which inhibits RNA and protein production) and withanolide D (which inhibits RNA production). The RNA and protein inhibition may lead to increased cancer cell death. *W. somnifera* has been found to elicit both antioxidant and pro-oxidant activities. Tumour-bearing animals treated with this drug showed increased GSH, SOD, glutathione peroxidase and catalase in the liver and skin. These effects could clearly repair oxidative damage caused by tumour growth and inflammation, thus reducing the likelihood of disease progression. This antioxidant activity is enhanced by the potential of *W. somnifera* to up-regulate phase II liver enzymes. *W. somnifera* may also mitigate unregulated cell growth via the potent tumour suppressor gene p53, which regulates cell cycle proliferation. *W. somnifera* was identified via mass spectrometry as the most potent constituent of *W. somnifera* to inhibit tumour necrosis factor-α induced NF-kappaB activation, inhibiting angiogenesis at a dose of 7 μg per kg per day. The NF-kappaB may play a key role in the antitumour action of *W. somnifera* since it is activated by carcinogens, tumour promoters and inflammatory agents. This implicates NF-kappaB suppression as one mechanism by which *W. somnifera* could decrease inflammation, enhance cytotoxicity and apoptosis of tumour, and decrease metastasis. *W. somnifera* also exerts a beneficial effect on immune system, which may explain some of its antitumour effects. The NK cell activity was significantly enhanced by *W. somnifera* during tumorigenesis. The strong immune-stimulating effect of *W. somnifera* elicits from macrophages and NK cells can increase tumour cell surveillance and control (Madhuri and Pandey, 2009c; Pandey *et al.*, 2013a).

Emblica officinalis as an Anticancer Drug

Various Uses of *Emblica officinalis*

Emblica officinalis Gaertn. (*Phyllanthus emblica* Linn., Amla, Indian gooseberry) belongs to Euphorbiaceae plant family. Amla fruits are acrid, cooling, refrigerant, astringent, diuretic and laxative. The raw fruits are aperient; while the dried fruits are useful in inflammation, haemorrhage, cough, diarrhoea and dysentery, and in combination with Fe, used for anaemia, jaundice and dyspepsia. The flowers of *E. officinalis* are cooling, refrigerant and aperient; while the root and bark are astringent. The fermented liquor prepared from the fruits is used in jaundice, dyspepsia and cough. Exudation from incision on the fruit is used as external application for the inflammation of eye. The seeds of *E. officinalis* are used for asthma, bronchitis and biliousness. Vitamin C (ascorbic acid or ascorbate), tannins (*e.g.*, emblicanins A and B) and flavonoids present in Amla have very powerful immunomodulatory, antioxidant and anticancer activities. Due to rich vitamin C, this herb is successfully used in the treatment of human scurvy. Quercetin present in amla has hepatoprotective effect. Phyllembin of Amla potentiates the pharmacological action of adrenaline both *in vitro* and *in vivo* (Madhuri *et al.*, 2011b; Pandey, 2011b; Pandey and Pandey, 2011; Pandey *et al.*, 2013a; Prajapati *et al.*, 2003).

E. officinalis is also used as an antipyretic, analgesic, cytoprotective, antitussive and gastroprotective agent. Additionally, it is useful in memory enhancing, ophthalmic disorders and lowering cholesterol level. It is also helpful in neutralizing snake venom, and used as an antimicrobial agent (Madhuri *et al.*, 2011b; Pandey *et al.*, 2013a).

Phytoconstituents of *Emblica officinalis*

E. officinalis fruit (Amla) contains ellagic acid, gallic acid, quercetin, kaempferol, flavonoids, glycosides, proanthocyanidins, vitamin C (ascorbic acid or ascorbate), and tannins (*e.g.,* emblicanins A and B). The fruits have also been reported to contain phyllemblic acid, lipid, emblicol, colloidal complexes, micic acid, amino acids and minerals. Phyllembin from fruit pulp identified as ethyl gallate; tannin from bark and leaves; fixed oil, essential oil and phosphatides from seeds; and leucodelphinidin from bark have also been isolated (Madhuri *et al.,* 2011b; Pandey, 2011b; Pandey and Pandey, 2011; Pandey *et al.,* 2013a; Prajapati *et al.,* 2003).

Anticancer and Related Activites of *Emblica officinalis*

Antioxidant Activity

Ethyl acetate extract of *E. officinalis* fruit has been reported to reduce the elevated levels of urea nitrogen and serum creatinine in aged rats. Oral administration of this extract significantly reduced the thiobarbituric acid-reactive substance levels of serum, renal homogenate and mitochondria in aged rats, suggesting that this drug would ameliorate the oxidative stress under aging. Increased inducible nitric oxide synthase and cyclooxygenase expression in the aorta of aging rats were also significantly suppressed by the ethyl acetate extract. This extract reduced the cyclooxygenase and nitric oxide synthase expression levels by inhibiting NF-kappaB activation in the aged rats. Thus, this herb would be a very useful antioxidant for the prevention of age-related renal disease (Yokozawa *et al.,* 2007). Pre-feeding of Amla appeared to reduce the hexachlorocyclohexane-enhanced activity of renal gamma-glutamyl transpeptidase (GGT). This shows the elevation of hepatic antioxidant system and lowering of cytotoxic products, which were otherwise affected by the administration of hexachlorocyclohexane (Anilakumar *et al.,* 2007). The effect of Amla extract on the oxidative stress in streptozotocin-induced diabetic rats was also noted. The orally administered extract of Amla showed strong free-radical scavenging activity. The extract slightly improved the body weight, and also significantly increased the various oxidative stress indices of the serum of diabetic rats. Moreover, the decreased levels of albumin in the diabetic rats were significantly improved with this drug. It also significantly improved the serum adiponectin levels. Thus, Amla can be used for relieving the oxidative stress and improving glucose metabolism in diabetes (Rao *et al.,* 2005).

E. officinalis is used to protect the skin from the devastating effects of free radicals, non-radicals and transition metal-induced oxidative stress. It is suitable for use in anti-ageing, general purpose skin care products and as sunscreen. *E. officinalis* fruits contain tannoid principles that have been reported to exhibit antioxidant activity *in vitro* and *in vivo*. Emblicanin-A (37 per cent) and emblicanin-B (33 per cent) enriched fractions of fresh juice of *E. officinalis* fruits were investigated for antioxidant activity against ischemia-reperfusion-induced oxidative stress in rat heart. The study confirmed the antioxidant effect of *E. officinalis,* and also indicated that the fruits of plant may exhibit a cardioprotective effect. The antioxidant activity of *E. officinalis* extract associated with the presence of hydrolysable tannins having ascorbic acid-like action have been also reported. *E. officinalis* contains tannoid principles comprising of emblicanin A, emblicanin B, punigluconin and pedunculagin, which have been reported to possess the antioxidant activity both *in vitro* and *in vivo*. Ellagic acid, as a powerful antioxidant present in *E. officinalis,* has the ability to inhibit mutations in genes and repairs the chromosomal abnormalities (Pandey, 2011b; Pandey *et al.*, 2013a).

Immunomodulatory Activity

Immune activation is an effective as well as protective approach against emerging infectious diseases. *E. officinalis* has been reported to inhibit the Cr-induced free radical production, and it restored the antioxidant status back to control level. It also inhibited the apoptosis and DNA fragmentation induced by Cr. It relieved the immunosuppressive effect of Cr on lymphocyte proliferation, and even restored the IL-2 and gamma-IFN production (Sai Ram *et al.*, 2003). *E. officinalis* and *Evolvulus alsinoides* (Shankhpushpi) were assessed for immunomodulatory activity in adjuvant induced arthritic rat model. Complete Freund's adjuvant was injected in right hind paw of the animals induced inflammation. Lymphocyte proliferation activity and histopathological severity of synovial hyperplasia were used to study the antiinflammatory response of both the extracts, which showed a marked reduction in inflammation and oedema, and caused immunosuppression in adjuvant induced arthritic rats, indicating that these drugs may provide an alternative approach for the treatment of arthritis (Ganju *et al.*, 2003). The immunomodulatory activity of the combined extracts of *O. sanctum*, *W. somnifera* and *E. officinalis* has also been observed (Madhuri, 2008; Pandey *et al.*, 2013a).

Anticancer Activity

E. officinalis has been reported to possess many medicinal properties, including immune-stimulator and antitumour activities (Jeena *et al.*, 2001). *E. officinalis* inhibits the growth and spread of various cancers, including breast, uterus, pancreas, stomach and liver cancers, and malignant ascites. It reduces the side effects of chemotherapy and radiotherapy. It reduces the cytotoxic effects in mice dosed with carcinogens (Pandey, 2011b). Amla contains 18 compounds that inhibit the growth of tumour cells, like gastric and uterine cancer cells (Zhang *et al.*, 2004). It enhanced NK cell activity in tumours. Its extract reduced the ascites and solid tumours induced by DLA cells in mice. The extract also increased the life span of tumour bearing animals. Emblicanins A and B (tannins) of *E. officinalis* possess strong antioxidant and anticancer properties (Pandey, 2011b). Suppression of the growth of cancer cells due to gallic acid, a major polyphenol as observed in Triphala, has also been reported (Kaur *et al.*, 2005).

Chemoprevention with food phytochemicals is presently considered as one of the most important strategies to control cancer. Chemopreventive potential of Amla extract on DMBA induced skin tumorigenesis in Swiss albino mice have been seen (Sancheti *et al.*, 2005). *E. officinalis* fruit administered in different concentrations (100, 250, 500 mg/kg, oral) for 7 consecutive days in Swiss albino mice prior to a single ip injection of DMBA, decreased the frequency of bone marrow micronuclei. The protection provided by *E. officinalis* fruit might be due to its antioxidant capacity and through its immunomodulatory effect on hepatic activation and detoxifying enzymes (Banu *et al.*, 2004). Phenolic compounds of Amla exhibit a number of beneficial effects, and can potentially inhibit several stages of carcinogenesis. Efficacy of the polyphenol fraction of *E. officinalis* on induction of apoptosis in mouse and human carcinoma cell lines, and its immunomodulatory effect on DENA induced liver tumours in rats was also obtained. The polyphenol fraction of *E. officinalis* could induce the apoptosis in DLA and CeHa cell lines. The polyphenol fraction also inhibited the DNA topoisomerase I in *Saccharomyces cervisiae*, mutant cell cultures and activity of cdc25 tyrosine phosphatase (Rajeshkumar *et al.*, 2003). Amla extract was found most active in inhibiting *in vitro* cell proliferation towards human tumour cell lines, including human erythromyeloid K562, T-lymphoid Jurkat, B-lymphoid Raji, erythroleukemic HeLa cell lines (Khan *et al.*, 2002).

Cyclophosphamide is one of the most famous alkylating anticancer drugs besides its toxic effects, including haematotoxicity, immunotoxicity and mutagenicity. *E. officinalis* may be beneficial as a component of combination therapy in cancer patients under cyclophosphamide treatment (Haque *et al.*, 2001). Phenolic compounds and major components from the fruit juice, branches, leaves and roots of *E. officinalis* showed stronger inhibition against B16F10 cell growth than against HeLa and MK-1 cell growth. Norsesquiterpenoid glycosides from the root showed significant antiproliferative activities (Zhang *et al.*, 2004). The cytoprotective and immunomodulating properties of *E. officinalis* against Cr(VI) induced oxidative damage have been reported. The Cr(VI) at the dose of 1 µg per ml caused severe cytotoxicity. It enhanced free radical production and decreased the reduced GSH levels and glutathione peroxidase activity in macrophages. However, Amla caused an enhanced cell survival, decreased free radical production and higher antioxidant levels. Further, Cr(VI) treatment resulted in decreased phagocytosis and gamma-interferon production; while Amla inhibited the Cr-induced immunosuppression, and restored both phagocytosis and gamma-interferon production by macrophages significantly (Pandey, 2011b; Pandey *et al.*, 2013a; Sai Ram *et al.*, 2003).

Silybum marianum in the Cancer

Silybum marianum in Various Diseases

Silybum marianum (Linn.) Gaertn. (milk thistle or bank thistle) belongs to the Compositae/Asteraceae plant family. The active principle, silymarin is a flavonolignan (polyphenolic fraction), extracted from the seeds (and may also from the fruits) of *S. marianum*. Silymarin and and its isomer silybin have been found to provide cytoprotection and above all, hepatoprotection (Kshirsagar *et al.*, 2009). Silymarin has been reported to treat various liver disorders as it has established the efficacy in restoration of liver function and regeneration of liver cells. It antagonized the toxin (α-aminitine) of *Amanita phalloides* and provided hepatoprotection against toxicity caused by phalloidine, galactosamine, paracetamol, carbon tetrachloride, thioacetamide and halothane (Kshirsagar *et al.*, 2009; Pandey, 1990; Pandey and Sahni, 2011b; Vogel, 1977). Silymarin has also protected the hepatocytes from injury due to poisoning, ischaemia, radiation, iron overload and viral hepatitis, so it is included in the pharmacopoeia of many countries, and is often used as supportive therapy in food poisoning by fungi, and in chronic liver disorders like steatosis and alcohol-related liver disease (Kshirsagar *et al.*, 2009; Pandey and Sahni, 2011b).

Furthermore, silymarin is used medicinally to treat liver disorders, including acute and chronic viral hepatitis, toxin/drug-induced hepatitis, cirrhosis and alcoholic liver diseases. It is also effective in certain cancers. Silymarin and its main component silybinin are used almost for hepatoprotection in human. Silymarin offers good protection in various toxic models of experimental liver diseases in laboratory animals. It possesses antioxidative, antiinflammatory, antifibrotic, antilipid peroxidative, membrane stabilizing and liver regenerating activities/mechanisms. Its clinical uses in humans comprise therapy in alcoholic liver diseases, liver cirrhosis, *Amanita* mushroom poisoning, viral hepatitis, toxic and drug-induced liver diseases (Pandey and Sahni, 2011b).

The hepatoprotective/hepatogenic activity of silymarin or extracts of *S. marianum* xenobiotic intoxication and fungal intoxication has been reported. Silymarin completely neutralized the hepatotoxic effect of various agents as evidenced by significant reduction in prolongation of hexobarbitol sleeping time and increased serum levels of transaminases and sorbitol dehydrogenase at the dose of 100 mg per kg, iv against carbon tetrachloride (0.15 ml/kg, oral) poisoning in rat. Similarly, a 100 per cent protection by silymarin (50 mg/kg, iv) against phalloidine (3 mg/kg, ip) hepatotoxicity and a marked hepatoprotective effect of silymarin (75 mg/kg, iv) in hepatotoxicity induced by α-aminitine (0.5 mg/kg, ip) in mouse were recorded. A significant reduction and restoration of the serum transaminase activity was seen after silymarin dosing during praseodymium and galactosamine induced hepatotoxicity (Kshirsagar *et al.*, 2009; Pandey, 1990; Pandey and Sahni, 2011b). The hepatogenic effects of the AqE and petroleum ether extract of *S. marianum* seeds were studied. These extracts at the dose of 1000 mg per kg body weight, orally, daily from 3rd to 7th day of the experiment produced beneficial results against paracetamol (500 mg/kg, orally, once on 1st day) induced hepatotoxicity. The AqE and PEE of *S. marianum* seeds significantly (P<0.05) improved the paracetamol altered activities of SAP, serum arginase, SGOT, SGPT and serum proteins. *S. marianum* seeds also caused regeneration of hepatic tissues in rats. The normalization and regeneration of liver tissues were also produced in mice (Pandey, 1990).

Silymarin has been shown to prevent carbon tetrachloride-induced lipid peroxidation and hepatotoxicity. Silibinin preserved the functional and structural integrity of hepatocyte membranes by preventing alterations of their phospholipid structure produced by carbon tetrachloride, and by restoring SAP and GGT activities. Histochemical and histoenzymological studies have shown that silymarin, administered 60 minutes before or no longer than 10 minutes after induction of acute intoxication with phalloidine, is able to neutralize the effects of the toxin and to modulate hepatocyte function. Similar results were obtained in dogs treated with sublethal oral doses of *A. phalloides,* in which hepatic injury was monitored by measuring enzymes and coagulation factors. Using the model of microcystin, which produced the acute hepatotoxicity in mice and rats, the neutralization of microcystin's lethal effects and pathological alterations by silymarin were also seen. The hepatoprotective activity of silymarin has also been demonstrated against partial hepatectomy models and toxic models in experimental animals after administration of acetaminophen (paracetamol), carbon tetrachloride, ethanol, galactosamine and

A. phalloides toxin. Silybin dihemisuccinate (a soluble form of silymarin) protected the rats against liver glutathione depletion and lipid peroxidation induced by acute acetaminophen hepatotoxicity and showed potential benefits of silymarin as an antidote. The hepatoprotective activity of silymarin against ethanol induced damage has been noticed in tested animals as evidenced from improvements in some LFTs, *e.g.*, SGOT, SGPT and GGT. Silymarin also protected liver tissues from injury due to ischaemia, radiation and viral hepatitis. It also protected against fumonisin B$_1$ (a mycotoxin produced by *Fusarium verticillioides* found on corn and corn-based foods) liver damage by inhibiting biological functions of free sphingoid bases and increasing cellular regeneration (Kshirsagar *et al.*, 2009; Pandey, 1990; Pandey and Sahni, 2011b).

Milk thistle plant has been reported to be used as antidiabetic, hepatoprotective, hypocholesterolaemic, antihypertensive, antiinflammatory, anticancer and antioxidant. Its seeds are also used as an antispasmodic, neuroprotective, antiviral, immunomodulant, cardioprotective, demulcent and antihaemorrhagic. This plant also serves as a galactagogue, and used in the treatment of uterine disorders (Kumar *et al.*, 2011). Overall, silymarin possesses antioxidant, immunomodulatory, anticancer, antiinflammatory, antihepatoxic and some other pharmacological activities. Its effectiveness against multiple disorders makes it a very promising drug of natural origin (Pandey and Sahni, 2011b). In conclusion, milk thistle seeds contain a powerful phytonutrient called, silymarin, which is increasingly being recognized for its wide-range of anticancer, antiinflammatory, antioxidative and protective effects against various drugs and toxins. Most studies point to silybin as the main active polyphenol compound in silymarin extracts. Milk thistle seeds have beneficial effects against various conditions, such as hepatitis (alcoholic, hepatitis B and C viruses), toxin-induced liver damage (drugs, industrial chemicals, liver-toxic chemotherapy regimens, poisonous '*death cap*' mushrooms, etc.), kidney damage (nephropathy) due to diabetes, cisplatin chemotherapy-induced nephropathy, diabetes, reduced insulin resistance, high blood sugar levels, '*bad*' LDL cholesterol and cancers of different organs. The supplement of milk thistle is available as a capsule, tablet, powder and liquid extract. The powdered milk thistle can be made into a tea. A typical daily dose ranges from 140 to 600 mg of silymarin, usually divided into 2 or 3 doses.

Phytoconstituents of *Silybum marianum*

Silymarin mainly contains three flavonolignan isomers, *i.e.*, silybin (or silibinin), silydianin and silychristin (Kshirsagar *et al.*, 2009; Pandey, 1990; Vogel, 1977). However, Kshirsagar *et al.* (2009) also elucidated that silymarin consists of four flavonolignan isomers, *viz.*, silybin, isosilybin, silydianin and silychristin with an empirical formula $C_{25}H_{22}O_{10}$, and the structural similarity of silymarin to steroid hormones is believed to be responsible for its protein synthesis facilitatory actions. Of all the isomers that constitute silymarin, silybin is the most active (Pandey and Sahni, 2011b). Another report (Agarwal *et al.*, 2006) states that silymarin consists of a family of flavonoids (silybin, isosilybin, silychristin, silydianin and taxifolin) commonly found in the dried fruit of the milk thistle (*S. marianum*) plant. Further studies revealed certain phytoconstituents of milk thistle, such as silybin A, silybin B, isosilybin A, isosilybin B, silychristin, silydianin, apigenin 7-O-β-(2'-O-α-rhamnosyl)-

galacturonide, kaempferol 3-O-α-rhamnoside-7-O-β-galacturonide, apigenin 7-O-β-glucuronide, apigenin 7-O-β-glucoside, apigenin 7-O-β-galactoside, kaempferol-3-O-α-rhamnoside, kaempferol, taxifolin and quercetin (Kumar *et al.*, 2011).

Anticancer and Related Activities of *Silybum marianum*

Milk thistle has anticancer activity against several cancer cell types, such as prostate cancer, breast, cervix, ovary, colon, lung, liver and skin cancers. Milk thistle has also been shown to enhance the efficacy of chemotherapy (cisplatin and doxorubicin) on certain cancers. Bhatia *et al.* (1999) treated different prostate, breast and cervical human carcinoma cells with silibinin, resulting in a highly significant inhibition of both cell growth and DNA synthesis in a time-dependent manner with large loss of cell viability only in case of cervical carcinoma cells. The higher doses of silymarin induced programmed cell death specifically in human ectocervical carcinoma A431 cells.

The extracts (commonly called silibinin and silymarin) from milk thistle seeds possess anticancer actions on human prostate carcinoma *in vitro* and *in vivo*. Seven distinct flavonolignan compounds and a flavonoid have been isolated from commercial silymarin extracts. Most notably, two pairs of diastereomers, silybin A and silybin B and isosilybin A and isosilybin B, are among these compounds. On the contrary, silibinin is composed only of a 1:1 mixture of silybin A and silybin B. Isosilybin B was the most potent suppressor of cancer cell growth relative to either the other pure constituents or the commercial extracts. Isosilybin A and isosilybin B were also the most effective suppressors of prostate-specific antigen secretion by androgen-dependent LNCaP cells. Silymarin and silibinin were shown for the first time to suppress the activity of DNA topoisomerase IIα gene promoter in DU145 cells and, among the pure compounds, isosilybin B was again the most effective. Thus, isosilybin B composes no more than 5 per cent of silymarin and is absent from silibinin; however, many other more abundant flavonolignans do ultimately influence the same end points at higher exposure concentrations. These results are indicate that extracts enriched for isosilybin B, or isosilybin B alone, might possess improved potency against prostate cancer (Paula *et al.*, 2005). Experimentally, UV radiation-induced immunosuppressive activity of silymarin in rodents was noted, and silibinin inhibited activation of human T-lymphocyte, human polymorphonuclear leucocyte. Silymarin also significantly suppressed inflammatory mediators, expression of histocompatibility complex molecules and nerve cell damage. Long-term dosing of sliymarin improved immunity by increasing T-lymphocytes, interleukins and reducing all types of immunoglobulins (Meeran *et al.*, 2006).

Silybin (pure, chemically defined substance) and silymarin (flavonoid complex) from *S. marianum* seeds were shown to have anticancer and canceroprotective, and also hypocholesterolemic activities. These effects were observed in the disorders of various organs, *e.g.*, prostate, lungs, CNS, kidneys, pancreas and skin. Besides the cytoprotective activity of silybin mediated by its antioxidative and radical-scavenging properties, also new functions based on the specific receptor interaction were discovered. These were studied on the molecular level, and modulation of various cell-signaling pathways with silybin was disclosed. Pro-apoptotic activity of silybin in pre- and/or cancerogenic cells and antiangiogenic activity of silybin are other important findings that bring silymarin preparations closer to respective application in the cancer treatment. Discovery of the inhibition and modulation of drug transporters, P-glycoproteins, ERs, nuclear receptors by silybin and some of its new derivatives contribute to better understanding of silybin activity on the molecular level. Silymarin application in veterinary medicine has also been observed. The works using optically pure silybin diastereomers clearly indicate extreme importance of the use of optically active silybin. Significance of silymarin and its components in the medicine have been clearly recorded (Gazak *et al.*, 2007).

Although role of silymarin as an antioxidant and hepatoprotective agent is well known, its role as an anticancer agent has begun to emerge. Extensive research within the last decade has shown that silymarin can suppress the proliferation of a variety of tumour cells (*e.g.*, prostate, breast, ovary, colon, lung, bladder). Many studies have shown that silymarin is a chemopreventive agent *in vivo* against a variety of carcinogens/tumour promoters, including UV light, DMBA, phorbol 12-myristate 13-acetate and others. Silymarin has also been shown to sensitize tumours to chemotherapeutic agents through down-regulation of the MDR protein and other mechanisms. It binds to both ER and androgen receptors, and down-regulates PSA. In addition to its chemopreventive effects, silymarin exhibits antitumour activity against human tumours (*e.g.*, prostate and ovary) in rodents. Clinical trials have shown that silymarin is bioavailable and pharmacologically safe (Agarwal *et al.*, 2006). A large number of studies have established the cancer chemopreventive role of silymarin in both *in vivo* and *in vitro* models. Silymarin elicited antiinflammatory as well as antimetastatic activity. The protective effects of silymarin and its major active constituent, silibinin, studied in various tissues, suggest a clinical application in cancer patients as an adjunct to estabilished therapies, to prevent or reduce chemotherapy as well as radiotherapy-induced toxicity (Ramasamy and Agarwal, 2008). The production of superoxide anion radicals and nitric oxide after treatment in the isolated rat Kupffer cells with silybin was inhibited. Treatment with silibinin (200 mg/kg) improved the liver steatosis and inflammation and decreased non-alcoholic steatohepatitis-induced lipid peroxidation, plasma insulin and plasma tumour necrosis factor-α. Silibinin also decreased the superoxide radical release and returned the relative liver weight as well as GSH back to normal (Pandey and Sahni, 2011b).

Mechanism of Action of *Silybum marianum* (Silymarin)

Suppression of cancer cells by silymarin is accomplished through the cell cycle arrest at the G_1/S-phase, induction of cyclin-dependent kinase inhibitors (*e.g.*, p15, p21 and p27), down-regulation of anti-apoptotic gene products (*e.g.*, Bcl-2 and Bcl-xL), inhibition of cell-survival kinases (AKT, PKC and MAPK) and inhibition of inflammatory transcription factors (*e.g.*, NF-kappaB). Silymarin can also down-regulate gene products involved in the proliferation of tumour cells (cyclin D1, EGFR, COX-2, TGF-β and IGF-IR), invasion (MMP-9), angiogenesis (VEGF) and metastasis (adhesion molecules). The antiinflammatory effects of silymarin are mediated through suppression of NF-kappaB-regulated gene products, including COX-2, LOX, inducible iNOS, TNF and IL-1 (Agarwal *et al.*, 2006). Silymarin modulates imbalance between cell survival and apoptosis through interference with the expressions of cell cycle regulators and proteins involved in apoptosis (Ramasamy and Agarwal, 2008).

Milk thistle has been investigated to protect against skin cancer in at least two ways: (a) it increases cell death in skin cells that have been damaged by UV radiation, which comprises 95 per cent of the rays from the sun; (b) it increases the repair of sun damaged skin cells from UV radiation, which comprises 5 per cent of the rays from the sun.

Silymarin has been reported to possess multiple mechanisms of actions. The antioxidant activity and cell regenerating functions as a result of increased protein synthesis are considered as most important actions. The ROS, *e.g.*, superoxide radical, hydroxyl radical, hydrogen peroxide and lipid peroxide radicals are produced as a normal consequence of biochemical processes in the body and as a result of increased exposure to xenobiotics. The mechanism of free radical damage includes ROS induced peroxidation of polyunsaturated fatty acid in the cell membrane bilayer, which causes a chain reaction of lipid peroxidation, thus damaging the cellular membrane and causing further oxidation of membrane lipids and proteins. Subsequently cell contents, including DNA, RNA and other cellular components are damaged. The cytoprotective effects of silymarin are mainly attributable to its antioxidant and free radical scavenging properties. Silymarin can also interact directly with cell membrane components to prevent any abnormalities in the content of lipid fraction responsible for maintaining normal fluidity.

The stimulation of protein synthesis is an important step in the repair of cell injury, and is essential for restoring the structural proteins and enzymes damaged by toxins. Overall, the cellular protection provided by silymarin appears to rest on four actions: (a) activity against lipid peroxidation as a result of free radical scavenging and the ability to increase the cellular content of GSH; (b) ability to regulate the membrane permeability and to increase membrane stability in the presence of xenobiotic damage; (c) capacity to regulate the nuclear expression by means of a steroid-like effect; and (d) inhibition of the transformation of stellate hepatocytes into myofibroblasts, which are collagen fibres leading to cirrhosis. Silymarin and silybinin inhibit the absorption of toxins, such as phalloidine or β-amanitine, preventing them from binding to the cell surface and inhibiting membrane transport systems. Further, silymarin and silibinin, by interacting with the lipid component of cell membranes, may influence their chemical and physical properties (Kshirsagar *et al.*, 2009; Pandey and Sahni, 2011b).

Well documented scavenging activity of silymarin and silibinin may explain the protection afforded by these substances against hepatotoxic agents. Silymarin and silybinin can exert their actions by acting as free radical scavengers and interrupting the lipid peroxidation processes involved in the hepatic injury produced by toxic agents. Both these are probably able to antagonise the depletion of the two main detoxifying mechanisms, GSH and SOD, by reducing the free radical load, increasing GSH levels and stimulating SOD activity. Silibinin probably acts not only on the cell membrane, but also on the nucleus, where it appeared to increase ribosomal protein synthesis by stimulating RNA polymerase-I and the transcription of rRNA. Silymarin works by acting as an antioxidant that prevents chain rupture. One of the mechanisms that can explain the capacity of silymarin to stimulate liver tissue regeneration is the increase in protein synthesis in the injured liver. In *in vivo* and *in vitro* experiments performed in the liver of rats from which part of the organ had been removed, silibinin produced a significant increase in the formation of ribosomes and in DNA synthesis, as well as an increase in protein synthesis. Silymarin can inhibit the hepatic cytochrome P450 detoxification system (phase I metabolism). It has been shown in mice that silibinin is able to inhibit numerous hepatic cytochrome P450 enzyme activities. This effect could explain some of the hepatoprotective activities of silymarin, especially against the intoxication due to *A. phalloides*. The *Amanita* toxin becomes lethal for hepatocytes only after having been activated by cytochrome P450 system. Inhibition of toxin bioactivation may contribute to limitation of its toxic effects. In addition, silymarin, together with other antioxidant agents, could contribute towards protection against free radicals generated by enzymes of cytochrome P450 system (Kshirsagar *et al.*, 2009; Pandey and Sahni, 2011b).

Immunostimulatory and Anticancer Effects of Medicinal Plants in Fishes

Role of Medicinal Plants in Various Diseases of Fish

Billions of dollars or rupees have been spent every year all over the world, but no perfect remedy of cancer could be investigated so far. Although, the medicinal plants (herbal drugs) have been with us for human therapy for millennia, there has been relatively little research on them to be used against fish diseases. Medicinal plants can be used not only as remedy, but also as growth promoters, stress resistance boosters and preventatives of infections in fishes. Thus, the medicinal plants in disease management are gaining success, because they are cheaper and exhibit no or very less toxicity. They are rich in a wide variety of phytochemicals, *e.g.*, tannins, alkaloids and flavonoids, which act against several diseases (Madhuri *et al.*, 2012b; Pandey and Madhuri, 2010e).

Studies have proved that herbal additives enhance the growth of fishes and protect them from diseases. The non-specific immune system of fish is considered to be the first line of defense against invading pathogens (Ahilan *et al.*, 2010). Inclusion of herbal additives in diets often provides cooperative action to various physiological functions. Beneficial role of vitamins C and E has been noted in fish nutrition, reproduction, growth and related indices. In addition, vitamins C and E are credited with modulating the stress response in fish. Biological role played by vitamins C and E is very vital for the sustained growth and health of many living organisms as well as fish. Dietary vitamins have antibody enhancement effects in fish. The synergistic effect of herbs has been reported in many fishes, including Japanese flounder and *Clarias gariepinus*. The beneficial utility of herbal growth promoters as an additive in the carp feed has been observed. There has been a significant difference between different herbal additives on the effect of growth rate in goldfish. The non-specific immune system of fish is the first line of defense against invading pathogens.

Neutrophils and phagocytes, lysozyme and complement are some important indices of non-specific immunity in fishes (Pandey *et al.*, 2012).

The medicinal plants can act as immunostimulants, conferring early activation to the non-specific defense mechanisms of fish and elevating the specific immune response. Medicinal plants have been used as medicine and an immune booster for humans for thousands of years. The herbs contain many immunologically active components such as polysaccharides, organic acids, alkaloids, glycosides and volatile oils, which can enhance immune functions. Therefore, the medicinal plants have been used as medicine to treat different fish diseases and to control of shrimp, especially in the countries like China, Mexico, India, Thailand and Japan (Yin *et al.*, 2008). There has been increased interest in the immune stimulating function of some herbs in aquaculture. The non-specific immune functions, such as bacteriolytic activity and leukocyte function were improved by some mixtures of Chinese herbs in shrimp (*Penaeus chinensis*) and tilapia (Chansue *et al.*, 2000). The non-specific defense mechanisms of fishes include neutrophil activation, production of peroxidase and oxidative radicals, together with initiation of other inflammatory factors (Ainsworth *et al.*, 1991).

Certain Plants with Immunostimulant and Anticancer Activities in Fish

Immunostimulatory activity of the AqE of *Eclipta alba* (Bhangra) leaf (oral administration as feed supplement) was studied in tilapia fish, *Oreochromis mossambicus*. The fishes were fed for 1, 2 or 3 weeks with diets containing the AqE of *E. alba* leaves at 0, 0.01, 0.1 or 1 per cent levels. After each week, non-specific humoral (lysozyme, antiprotease and complement) response, cellular (myeloperoxidase content, production of reactive oxygen and nitrogen species) response and disease resistance against *Aeromonas hydrophila* (a bacterial pathogen) were determined. The results indicated that the AqE of *E. alba* administered as feed supplement significantly enhanced most of the non-specific immune parameters tested. Among the humoral responses, lysozyme activity significantly increased after feeding with AqE for 1, 2 or 3 weeks. No significant modulation was noticed in all the cellular responses tested after 3 weeks of feeding, while the ROS production and myeloperoxidase content showed significant enhancement after 1 week of feeding with AqE. When challenged with *A. hydrophila* after 1, 2 or 3 weeks of feeding, the percentage mortality was significantly reduced in the treated fish. The highest dose of 1 per cent gave better protection than the other doses with the relative percentage survival values of 64, 75 and 32 after feeding for 1, 2 and 3 weeks respectively. Finally, the results indicated that dietary intake of AqE of *E. alba* leaves enhances the non-specific immune responses and disease resistance of *O. mossambicus* against *A. hydrophila* bacteria (Christybapita *et al.*, 2007).

Lonicera japonica herb has been known as an antiinflammatory agent and used widely for upper respiratory tract infections, diabetes mellitus and rheumatoid arthritis. It has been reported that *L. japonica* significantly increased the blood neutrophil activity and promoted phagocytosis by the neutrophils in bovine at the correct concentration (Lee *et al.*, 1998). After challenge with *A. hydrophila*, survival of fish fed with the extracts of *Ganoderma lucidum* and *L. japonica* herbs was improved.

The survival was further enhanced in the group fed with *Ganoderma* extract supplement and when both herbs were used together. It is possible that this is the result of enhancement of some components of non-specific immune system of the fish by *Ganoderma* and a combination of *Ganoderma* and *Lonicera*. There is strong evidence that feeding glucans can modify the activity of the innate immune system of fish and increase the disease resistance in several fish species (Madhuri *et al.*, 2012b). It has been shown that an AqE of *G. lucidum* promoted the phagocytosis by macrophages in mice immunosuppressed by an anticancer drug, cyclophosphamide. *G. lucidum* stimulated the proliferation of lymphocytes induced by concanavalin A or lipopolysaccharide and influenced the gene expression of cytokines (Wang *et al.*, 2003). Several studies have shown that the oral administration of chitin and yeast products (MacroGard, Vitastim and *Saccharomyces cerevisiae*) increased the phagocytic capability of the cells in rainbow trout. The extracellular activity was very high in fish fed with dietary glucan (Jeney *et al.*, 1997).

The non-specific immune effects of two Chinese herbs (*L. japonica* and *G. lucidum*) were determined on tilapia fishes. The herbal diets used were 1 per cent of *Lonicera*, 1 per cent of *Ganoderma* and a mixture of *Ganoderma* (0.5 per cent) and *Lonicera* (0.5 per cent). The diets were fed for 3 weeks. The respiratory burst activity of WBC, phagocytosis, plasma lysozyme, total protein and total immunoglobulin were monitored. Following 3 weeks after feeding, the fishes were infected with *A. hydrophila* and mortalities recorded. The study showed that feeding tilapia fishes with *Ganoderma* and *Lonicera* alone or in combination enhanced the phagocytosis by blood phagocytic cells during the whole experimental period and stimulated lysozyme activity after 2 weeks. Respiratory burst activity of phagocytic blood cells, total protein and total immunoglobulin in plasma were not enhanced. Both herbs when used alone or in combination increased the survival of fishes after challenge with *A. hydrophila*. The highest mortality (58 per cent) was observed in control fishes, followed with fishes fed with *Lonicera* extract (43 per cent) and fishes fed with *Ganoderma* (30 per cent). The lowest mortality (21 per cent) was observed when fishes were fed with a combination of these two medicinal plants. Hence, it can be concluded that the herbal extracts added to diets acted as immunostimulants, and appeared to improve the immune status and disease resistance of fish. It was also shown that *Astragalus* enhanced lysozyme activities in tilapia fishes during the whole period of the experiment when fed with low (0.1 per cent) and medium (0.5 per cent) doses. In case of fishes fed with *Scutellaria*, there was significant inhibition of extracellular superoxide anion production (Yin *et al.*, 2008).

The anticancer effect of *Solanum nigrum* (Makoi, black nightshade) herb was observed by Patel *et al.* (2009) in fish. They noted the significant cytotoxic effect of *S. nigrum* fruit methanolic extract (10-0.0196 mg/ml) on HeLa cell lines of fish. Efficacy of the polyphenol fraction of *Emblica officinalis* (Amla) fruit on the induction of apoptosis in mouse and human carcinoma cell lines, and its immunomodulatory effect on DEN induced liver tumours in rats was found. The polyphenol fraction of Amla induced the apoptosis in DLA and CeHa cell lines (Madhuri *et al.*, 2012b; Rajeshkumar *et al.*, 2003).

Yin *et al.* (2008) reported that the oral administration of ginger (*Zingiber officinale*) extract increases the phagocytic capability of cells in rainbow trout; while the extracts of four Chinese herbs (*Rheum officinale, Andrographis paniculata, Isatis indigotica* and *Lonicera japonica*) increased the phagocytosis of WBCs of crucian carp. Ahilan *et al.* (2010) observed that the addition of *Phyllanthus niruri* and *Aloe vera* (Aloe) as herbal additives can positively enhance the growth performance of goldfish, *Carassius auratus*. The medicinal plants, *viz.*, ginger, nettle and mistletoe as an adjuvant therapy in rainbow trout through feed enhanced the phagocytosis, and cellular and humoral defense mechanisms against pathogens. The disease resistant of *Catla catla* fish was produced through immersion treatment of three herbs, *viz.*, *Allium sativum* (Garlic), *Azadirachta indica* (Neem) and *Curcuma longa* (Haldi rhizome, turmeric) in spawn. Nargis *et al.* (2011) have seen the immunostimulant effects of the dietary intake of *A. sativum* and *Vitex negundo* extracts on the *Labeo rohita* (Indian major carp) fingerlings.

The immunostimulant effects of the dietary intake of various medicinal plant extracts on fish, rainbow trout (*Oncorhynchus mykiss*), were investigated. The fishes were fed with diets containing AqEs of mistletoe (*Viscum album*), nettle (*Urtica dioica*) and ginger (*Zingiber officinale*). The food containing lyophilized extracts of these plants as 0.1 per cent and 1 per cent was used at a rate of 2 per cent of body weight per day for 3 weeks. The plant materials tested for immunostimulatory food additives caused an enhanced extracellular respiratory burst activity ($P<0.001$). The fishes fed with a diet containing 1 per cent AqE of powdered ginger roots exhibited a significant non-specific immune response. Phagocytosis and extracellular burst activity of blood leukocytes (WBCs) were significantly higher in this group than those in the control group. All plant extracts added to fish diet increased the total protein level in plasma, except 0.1 per cent ginger. The highest level of plasma proteins was observed in the group fed with 1 per cent ginger extract containing feed. It was shown that in trout fed with nettle and mistletoe extracts the production of extracellular superoxide anion was of a similar level to that in the control fish (Dugenci *et al.*, 2003).

In *A. hydrophila* infected goldfish (*Carassius auratus*), fed with diets containing 100 and 200 mg kg^{-1}) of mixed herbal extracts supplementation feeds, the WBC levels significantly increased ($P<0.05$). The RBC and Hb in goldfish significantly decreased ($P<0.05$) when fed with 100 and 200 mg kg^{-1} of mixed herbal extracts supplementation feeds; while it was restored near control when infected fish fed with 400 or 800 mg kg^{-1} of herbal extracts supplementation feeds. The haematocrit values declined significantly ($P<0.05$) in 100, 200 and 400 mg kg^{-1} of mixed herbal supplementation feeding groups on weeks 2 and 4 when compared to control group. The mean corpuscular volume, mean corpuscular Hb and mean corpuscular Hb concentration values almost significantly altered from the control values. The infected goldfish treated with 100 or 200 mg kg^{-1} of herbal supplementation feeds exhibited significant ($P<0.05$) decrease in total protein, glucose and cholesterol levels on week 1 to 4; whereas, it was restored when infected fish fed with 400 or 800 mg kg^{-1} of herbal supplementation feeds on week 4. In comparison to untreated control goldfish, the respiratory burst activity and phagocytic activity of blood cells was significantly enhanced in infected fish feeding with 200, 400 and 800 mg kg^{-1} of herbal supplementation feeds compared to the control. However, the infected fish fed with all the doses of mixed herbal supplementation feeds, the lysozyme activity was significantly enhanced throughout the experimental period. This study showed that the infected goldfish treated with 400 and 800 mg kg^{-1} of herbal supplementation feeds preceding the challenge with live *A. hydrophila* had 30 per cent and 25 per cent mortality. Whilst, 100 and 200 mg kg^{-1} of herbal supplementation feeds treated groups were found the mortalities of 50 per cent and 45 per cent, respectively. The results indicated that 400 or 800 mg kg^{-1} of mixed herbal supplementation feeds restored the altered hematological parameters and triggered the innate immune system of goldfish against *A. hydrophila* (Harikrishnan *et al.*, 2010).

Efficacy of dietary doses of *Withania somnifera* root powder was evaluated on immunological parameters and disease resistance against *A. hydrophila* infection in *L. rohita* fingerlings. These fishes were fed with dry diet containing *W. somnifera* root powder for 42 days. The immunological (NBT level, phagocytic activity, total immunoglobulin and lysozyme activity) parameters of fishes were examined at 0, 14, 28 and 42 days of feeding. The fishes were challenged with *A. hydrophila* for 42 days post-feeding and the mortalities (per cent) were recorded over 14 days post-infection. The results demonstrated that fishes fed with *W. somnifera* root showed enhanced NBT level, phagocytic activity, total immunoglobulin level and lysozyme activity (p<0.05) compared with the control group. The survivability was higher in experimental diets than the control group. The dietary *W. somnifera* showed significantly (P<0.05) higher protection against *A. hydrophila* infection. The results suggested that the *W. somnifera* root powder has a stimulatory effect on immunological parameters and increases disease resistance in *L. rohita* fingerlings against *A. hydrophila* infection (Sharma *et al.*, 2010).

Chapter 7
Anticancer Plant-Drugs: Part IV

A Herbal Drug Effective against Estrogen Induced Ovary Cancer

An estrogen, EO (@ 250, 500 and 750 µg/kg, orally, weekly for 8 and 12 weeks) induced hepatotoxicity (Pandey *et al.*, 2008) and uterine cytotoxicity (Madhuri *et al.*, 2009) were observed in female albino rats. Many herbal drugs are prevalent for the treatment of cancer, and ProImmu (a herbal formulation) has been reported to possess immunomodulatory effect and restore normal histoarchitecture of damaged tissues. In view of this fact, a study was undertaken (Madhuri and Pandey, 2012e), as mentioned here, to evaluate the anticancer effect of ProImmu on EO (estrogen) induced ovarian cancer/adenocarcinoma in rats by conducting biochemical and histopathological studies.

Experimental Study

Thirty healthy inbred female albino rats (each weighing 100-150 g) were divided into five groups (each group consisted of 6 rats). The animals were kept under standard laboratory conditions in animal house and fed on standard pellet diet and clean drinking water. However, the animals were fasted overnight before the experiment, but water was given *ad libitum*. The experimental designs and protocol in the study received the approval of IAEC. The lynoral tablets (containing 0.05 mg of EO only in each tablet) were triturated and suspended in distilled water mixed with a pinch of *Gum acacia* powder (since the drug is insoluble in water). The normal saline administered to the rats was also mixed with a pinch of *Gum acacia* powder to have uniformity with EO suspension. Other required chemicals and reagents were purchased from the chemical shops.

The rats of group 1 (normal) were administered normal saline; while the rats of groups 2 to 5 were administered EO at the dose rate of 750 µg per kg, orally, weekly for

24 weeks. However, the aqueous suspension of ProImmu (mixed with a pinch of *Gum acacia* powder, since ProImmu is insoluble in water) was administered at the dose rate of 500 mg per kg, orally, daily for 4, 8 and 12 weeks after 20, 16 and 12 weeks of EO administration in groups 3, 4 and 5, respectively. After experiment, the biochemical and histopathological studies were done.

For biochemical study, the blood of all the rats was collected from their eye veins and the serum was separated. Then, the activities (IU/L) of SGPT (Bradley *et al.*, 1972) and SAP (Wilkinson *et al.*, 1969) were estimated on computerized serum auto-analyzer. The biochemical data of SGPT and SAP were analyzed statistically, and the mean values and standard error (SE) were calculated. To find out the significance of difference among different groups of treatments, Duncan's new multiple range test at P=0.05 (5 per cent level of significance) in completely random design was employed (Steel and Torrie, 1980).

For histopathological study, the rats were sacrificed by cervical dislocation (euthanized scientifically) after collection of the blood. The ovaries of all the rats/groups were collected and preserved separately in 10 per cent buffered formalin. The ovarian tissues were processed and stained with Harris's H&E stain as per the method described by Culling (1963). On microscopic examination, the cancerous and anticancerous changes or normal histological profiles in the ovarian tissues were assessed.

The average values of SGPT and SAP obtained in groups 1 to 5 are presented in Table 12. The normal (Group 1) SGPT activity increased significantly (P<0.05) in groups 2 to 4, while it did not increase in group 5 and significantly returned to normal level on the 25th week of experiment. On the contrary, the SAP activity of group 1 (normal) significantly (P<0.05) increased in groups 2 to 5 on the 25th week of experiment. The SGPT and SAP activities of groups 2 to 5 differed significantly (P<0.05) with each other. On biochemical estimation, the increased SGPT activity recorded in the study may be correlated with the reports of Bradley *et al.* (1972) and Somkumar (2003), who stated the enhanced activity during various types of cancer. The markedly increased SAP activity has also been recorded in estrogen induced human endometrial cancer (Holinka *et al.*, 1986). The EO induced SAP activity of group 2 significantly reduced by ProImmu in groups 3 to 5, and it declined insignificantly towards the normal level. ProImmu improved SGPT activity was also seen by Das *et al.* (2000). The normalization in SGPT and SAP activities was also seen by Somkumar (2003) after administration of *O. sanctum* and *W. somnifera* (both plants are ingredients of ProImmu) in mice bearing S180 tumour.

The ovarian tissues of group 2 revealed marked fibrous tissue proliferation of follicular epithelium when compared with the tissues of group 1 (normal). The hyperplastic follicular epithelium at certain places revealed necrotic changes. Various cancerous lesions, including hyperchormatosis, enlargement of nuclei, anisokaryosis, anisocytosis and proliferation of blood vessels filled with RBCs were observed (Figure 16, Chapter 2). The changes were suggestive of the development of cancer (adenocarcinoma) in the ovary. In group 3, mild histopathological changes, including degeneration and necrosis were seen; however, the regeneration of some ovarian tissues was also noticed (Figure 17). In group 4, the ovarian tissues revealed more regeneration and improvement (Figure 18), and in group 5, the regeneration and normalization of damaged ovarian tissues in most of the areas were observed (Figure 19). Ishimura *et al.* (1986) also observed the collateral findings by reporting that the tumours arising from interstitial cell hyperplasia, with proliferation and invasion of the ovarian surface epithelium into ovarian stroma was caused by oestradiol (an estrogen). The ovarian surface epithelium cell proliferation was also observed by Murdoch and Van Kirk (2002) in sheep exposed with oestradiol. The ovarian cancer due to excessive and long-term use of oestrogen has been explained by Madhuri (2008), and Pandey and Madhuri (2008a). The cytogenic and regenerative effects of ProImmu observed in groups 3 to 5 resemble with the reports of many workers (Agrawala *et al.*, 2001; Nemmani *et al.*, 2002).

Table 12: Anticancer Effect of ProImmu on SGPT and SAP during EO Induced Ovarian Cancer in Albino Rat

Group*	Treatment	Week of Experiment	SGPT Activity (IU/L) Mean#±SE	SAP Activity (IU/L) Mean#±SE
1	Normal saline	1st	31.7d±1.0	171.5e ±1.8
2	EO @ 750 µg/kg, oral, once a week for 24 weeks	25th	80.3a±0.6	287.5a±1.6
3	EO @ 750 µg/kg, oral, once a week for 24 weeks + ProImmu @ 500 mg/kg, oral, daily for 4 weeks after 20 weeks of EO administration	25th	71.5b±1.5	237.7b±1.3
4	EO @ 750 µg/kg, oral, once a week for 24 weeks + ProImmu @ 500 mg/kg, oral, daily for 8 weeks after 16 weeks of EO administration	25th	53.4c±1.1	189.7c±1.1
5	EO @ 750 µg/kg, oral, once a week for 24 weeks + ProImmu @ 500 mg/kg, oral, daily for 12 weeks after 12 weeks of EO administration	25th	34.8d±1.1	177.7d±1.0

* Number of animals in each group=6.

Mean with same superscript does not differ significantly.

The beneficial effect of ProImmu against EO induced ovarian cancer observed in groups 3 to 5 may be further correlated with the findings of other investigators, who reported that the ingredients of ProImmu have anticancer activities. The anticancer activity of *O. sanctum* and *W. somnifera* was observed in mice bearing tumours (Somkumar, 2003). *W. somnifera* and *E. officinalis* showed inhibitory effects on hamster ovary cell lines (Sumanthran *et al.*, 2007). Conclusively, ProImmu caused regeneration and normalcy to a great extent against EO induced ovarian cancer. The anticancer effect of ProImmu may be presumably due to its (or its ingredients) immunostimulatory, antioxidant, phagocytic and other tissue protective activities (Madhuri, 2008; Madhuri and Pandey, 2012e; Pandey *et al.*, 2013a).

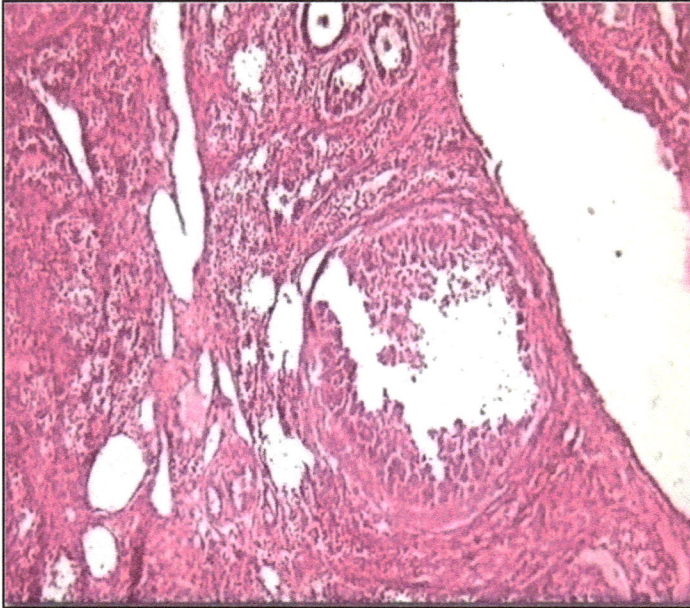

Figure 17: Photomicrograph (x100, H&E) of Rat Ovarian Tissues on 25th Week of EO (750 µg/kg, oral, weekly for 24 weeks) and Prolmmu (500 mg/kg, oral, daily for 4 weeks after 20 weeks of EO) Administration, Shows Mild Pathological Changes Including Degeneration and Necrosis; Regeneration in some Ovarian Tissues also Appear [Madhuri, 2008; Madhuri and Pandey, 2012e].

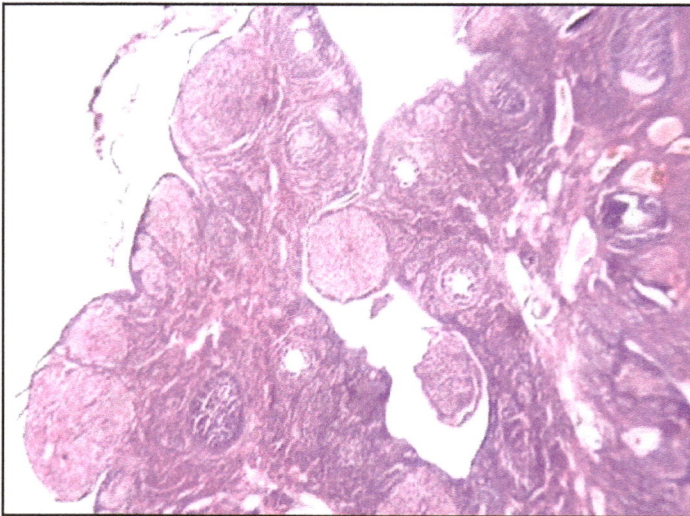

Figure 18: Photomicrograph (x100, H&E) of Rat Ovarian Tissues on 25th Week of EO (750 µg/kg, oral, weekly for 24 weeks) and Prolmmu (500 mg/kg, oral, daily for 8 weeks after 16 weeks of EO) Administration, Shows Reappearance of Normal Ovarian Tissues at many Places; Degeneration and Necrosis also Appear at some Places [Madhuri, 2008; Madhuri and Pandey, 2012e].

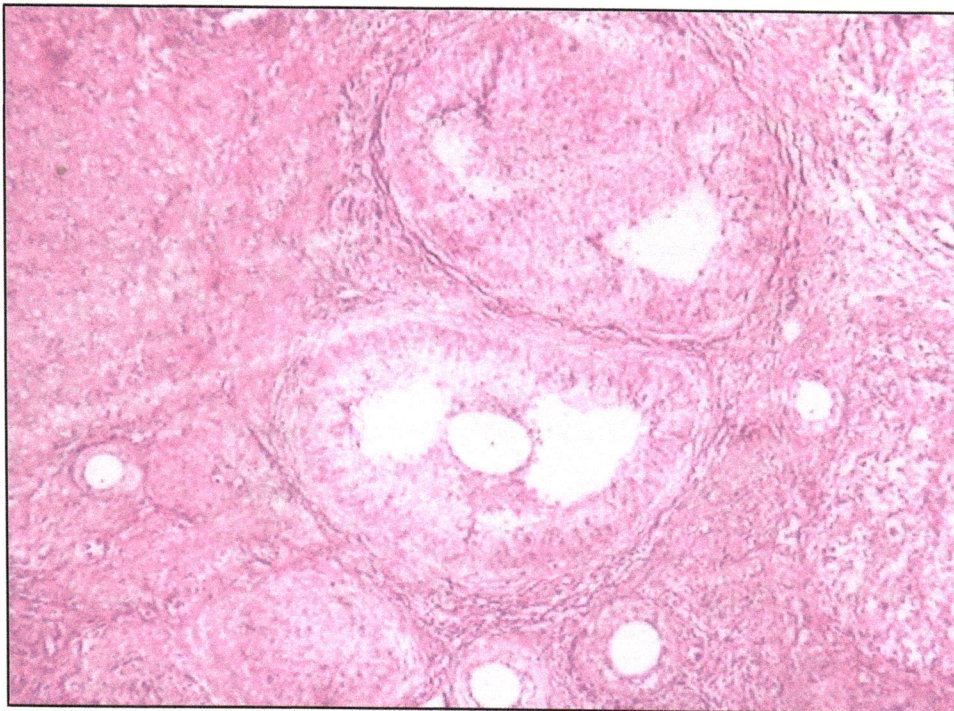

Figure 19: Photomicrograph (x100, H&E) of Rat Ovarian Tissues on 25th Week of EO (750 µg/kg, oral, weekly for 24 weeks) and ProImmu (500 mg/kg, oral, daily for 12 weeks after 12 weeks of EO) Administration, Shows much more Improvement and Regeneration [Madhuri, 2008; Madhuri and Pandey, 2012e].

Anticancer Effect of ProImmu on Estrogen Caused Uterus Cancer

The female hormone, estrogen causes cancers of many organs, including uterus (Madhuri, 2008; Pandey *et al.*, 2010). Many drugs are prevalent for treatment of cancer, but there is no perfect remedy and a plenty of research is still carrying all over the world. Some herbal products have been prepared from certain medicinal plants. Consequently, ProImmu has been reported to have immunomodulatory and genoprotective effects (Madhuri, 2008; Nemmani *et al.*, 2002). Hence, the study was performed to evaluate the anticancer effect of ProImmu on EO (estrogen) caused uterine cancer in albino rats on the basis of histopathological study of uterine tissues (Madhuri *et al.*, 2010).

Experimental Study

The experiment was conducted similar to the experimental designs and protocols mentioned under '*A herbal drug effective against estrogen induced ovary cancer*' (above experiment). For histopathological study, the rats were sacrificed by cervical dislocation (euthanized scientifically). The uteri of all the rats/groups were collected and preserved in 10 per cent buffered formalin. The uterine tissues were processed and stained with H&E stain as per the method of Culling (1963), and the uterine tissues were examined, microscopically for cancerous and anticancerous changes or normal histological profiles.

On the 25th week of experiment, the uterine tissues damaged by EO (Group 2) revealed the fibroblastic bundles made up of mature fibrocytes (as collagens). Focal hyperplasia of endometrial lining (epithelium) was quite evident. Other cancerous changes like hyperchromasia, enlargement of nuclei, anisokaryosis, anisocytosis, angiogenesis and glandular polarity or no glandular structures were observed. The smooth muscles were reactive and damaged. Disarray, leading to severe malignancy in the whole area was also seen (Figure 13, Chapter 2). The uterine tissues of group 3 (administered with EO for 24 weeks and ProImmu for 4 weeks) revealed mild histopathological changes including vacuolar degeneration, necrosis and fibrosis; however, the tissue regeneration in some areas was also observed (Figure 20). In group 4 (administered with EO for 24 weeks and ProImmu for 8 weeks), the necrobiotic changes were rather inconspicuous and reappearance of several normal tissues were seen (Figure 21). In group 5 (administered with EO for 24 weeks and ProImmu for 12 weeks), much better signs of improvement as evident from the reappearance of virtually normal histological profiles in the uterine tissues were noticed (Figure 22).

The histopathological changes, leading to uterine cancer (Figure 13) produced by EO may be correlated with the report of Meissner *et al.* (1957), who noticed the uterine cancer after administration of stilbestrol estrogen. Newbold and Liehr (2000) observed the uterine adenocarcinoma at 12 and 18 months after administration of EO and other estrogens in mice. ProImmu has been reported to cause restoration of normal tissues by increasing NK cell activity and proliferation of splenic leucocyte against K562 cells in mice. Furthermore, as mentioned earlier in Chapters 3 and 6, ProImmu and all of its four ingredients of ProImmu (*viz.*, *E. officinalis*, *O. sanctum*, *T. cordifolia* and *W. somnifera*) have anticancer, antioxidant and immunostimulatory effects (Agrawala *et al.*, 2001; Das *et al.*, 2000; Madhuri, 2008; Madhuri and Pandey, 2010a; Nemmani *et al.*, 2002). It may be, therefore, concluded that ProImmu has exhibited anticancer effect (as appear in Figures 20-22) against EO induced uterine cancer, which might be due to its immunostimulatory, antioxidant, phagocytic and its other tissue protective activities (Madhuri *et al.*, 2010).

Improvements of Biochemical Profiles by ProImmu during Uterine and Ovarian Cancer

A study was performed (Madhuri, 2008; Madhuri *et al.*, 2010), as described below, to evaluate the beneficial (*i.e.*, anticancer) effect of ProImmu on EO (oestrogen) induced uterine and ovarian cancer in rats by estimating the activities of certain biochemical parameters (LFTs).

Experimental Study

The experiment was conducted similar to the experimental designs and protocols mentioned under the experiment, '*A herbal drug effective against estrogen induced ovary cancer*'. For biochemical study, the blood of all the rats was collected from their eye veins and the serum was separated. The activities of SGOT (Bradley *et al.*, 1972),

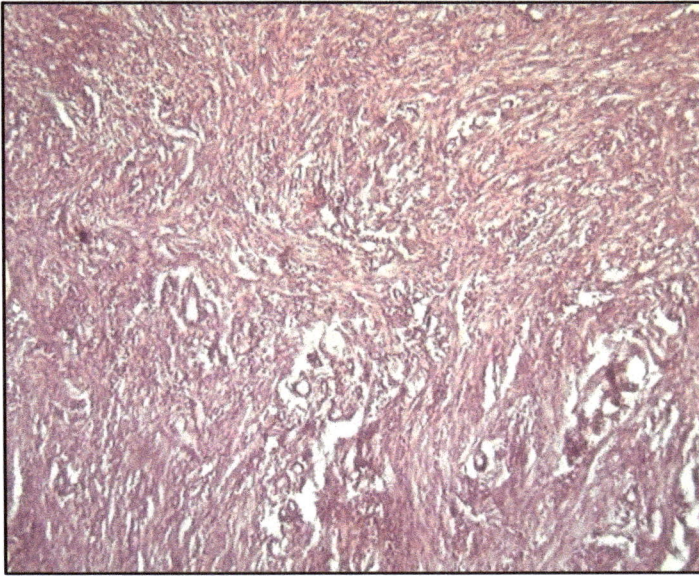

Figure 20: Photomicrograph (x100, H&E) of Rat Uterine Tissues on 25 th Week of EO (750 µg/kg, oral, weekly for 24 weeks) and ProImmu (500 mg/kg, oral, Daily for 4 Weeks after 20 Weeks of EO) Administration, shows Mild Histopathological Changes as Vacuolar Degeneration, Necrosis and Fibrosis with Regeneration of Uterine Tissues in some Areas [Madhuri, 2008; Madhuri *et al.,* 2010].

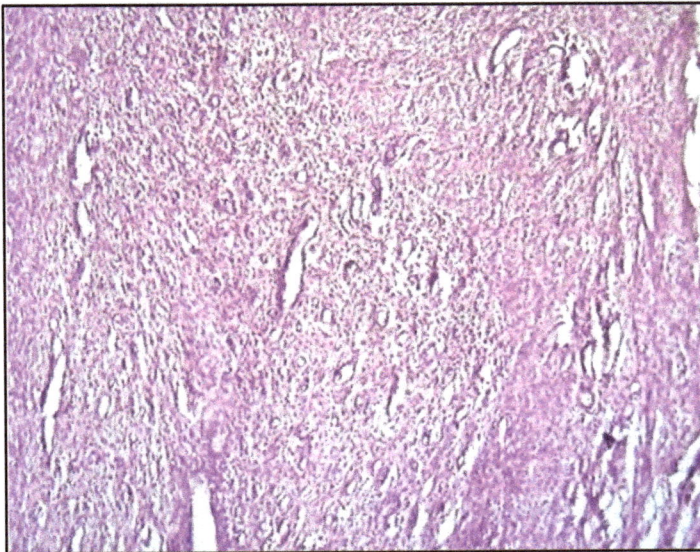

Figure 21: Photomicrograph (x100, H&E) of Rat Uterine Tissues on 25 th Week of EO (750 µg/kg, oral, weekly for 24 weeks) and ProImmu (500 mg/kg, oral, daily for 8 weeks after 16 weeks of EO) Administration, shows better Improvement with Reappearance of Several Normal Tissues [Madhuri, 2008; Madhuri *et al.,* 2010].

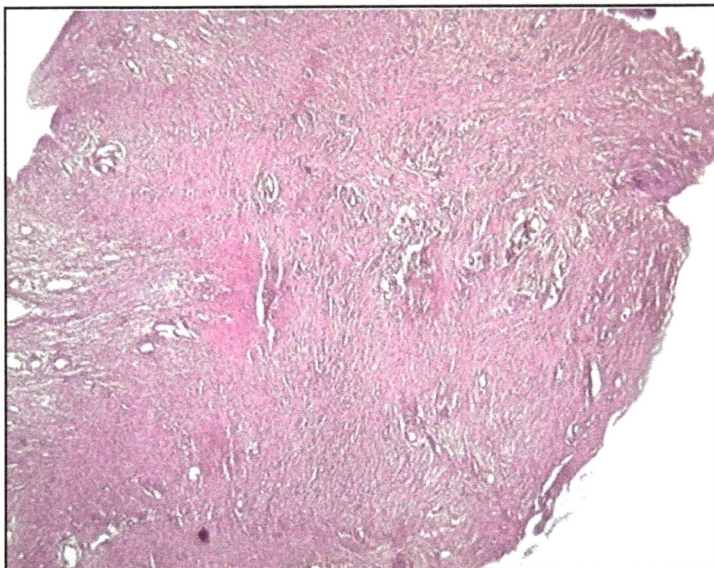

Figure 22: Photomicrograph (x100, H&E) of Rat Uterine Tissues on 25th Week of EO (750 µg/kg, oral, weekly for 24 weeks) and ProImmu (500 mg/kg, oral, daily for 12 weeks after 12 weeks of EO) Administration, shows much better Improvement with Reappearance of Virtually Normal Histological Profile [Madhuri, 2008; Madhuri *et al.*, 2010].

serum lactate dehydrogenase (SLDH; Bergmeyer, 1971), G6PDH (Kachmar and Moss, 1976) and serum total protein (STP; Flack and Woollen, 1984) were estimated on computerized serum auto-analyzer. The biochemical data of each parameter (LFT) were analyzed statistically, and the mean values and standard error (SE) were calculated. To find out the significance of difference among different groups of treatments, Duncan's new multiple range test at P=0.05 (5 per cent level of significance) in completely random design was employed (Steel and Torrie, 1980).

The mean±SE values (IU/L) of SGOT and SLDH estimated in different groups of rats are depicted in Table 13. The activities of two serum enzymes, *viz.*, SGOT and SLDH increased significantly (P<0.05) in groups 2 to 4 as compared to group 1 (normal) of rats. Further, the increased SGOT and SLDH activities of group 2 significantly decreased in groups 3 to 5. However, the SGOT activities of groups 3 and 4 did not differ with each other, but the SLDH activities of groups 3, 4 and 5 were significantly differed with each other. On the contrary, the values of groups 1 and 5 in relation to SGOT and SLDH did not differ significantly with each other.

The results of G6PDH and STP are shown in Table 14. The levels of G6PDH (U/g Hb) and STP (g/dl) in the rats of group 1 (normal) decreased significantly (P<0.05) in groups 2 to 4 on the 25th week of experiment. The decreased level of G6PDH enzyme in group 2 increased significantly in groups 3 to 5; whereas, the decreased levels of STP in groups 2 and 3 increased significantly in groups 4 and 5. However, there was no significant difference between the G6PDH activities of groups 1 and 5.

There was also non-significant difference between the STP activities of groups 1 and 5, and between groups 2 and 3.

Table 13: Effect of ProImmu on SGOT and SLDH during EO Induced Uterine and Ovarian Cancer in Albino Rat

Group*	Treatment	Week of Experiment	SGOT Activity (IU/L) Mean#±SE	SLDH Activity (IU/L) Mean#±SE
1	Normal saline	1st	77.4c±2.0	177.8d±3.4
2	EO @ 750 µg/kg, oral, once a week for 24 weeks	25th	123.2a±2.7	275.7a±2.1
3	EO @ 750 µg/kg, oral, once a week for 24 weeks + ProImmu @ 500 mg/kg, oral, daily for 4 weeks after 20 weeks of EO administration	25th	108.5b±4.6	232.0b±2.5
4	EO @ 750 µg/kg, oral, once a week for 24 weeks + ProImmu @ 500 mg/kg, oral, daily for 8 weeks after 16 weeks of EO administration	25th	105.1b±4.2	201.3c±1.7
5	EO @ 750 µg/kg, oral, once a week for 24 weeks + ProImmu @ 500 mg/kg, oral, daily for 12 weeks after 12 weeks of EO administration	25th	82.3c±0.8	183.5d±2.4

Table 14: Effect of ProImmu on G6PDH and STP during EO Induced Uterine and Ovarian Cancer in Albino Rat

Group*	Treatment	Week of Experiment	G6PDH Activity (U/g Hb) Mean#±SE	STP Activity (g/dl) Mean#±SE
1	Normal saline	1st	12.1a±0.2	5.98a±0.25
2	EO @ 750 µg/kg, oral, once a week for 24 weeks	25th	3.8d±0.1	4.02c±0.04
3	EO @ 750 µg/kg, oral, once a week for 24 weeks + ProImmu @ 500 mg/kg, oral, daily for 4 weeks after 20 weeks of EO administration	25th	5.5c±0.0	4.23c±0.03
4	EO @ 750 µg/kg, oral, once a week for 24 weeks + ProImmu @ 500 mg/kg, oral, daily for 8 weeks after 16 weeks of EO administration	25th	10.5b±0.1	4.77b±0.09
5	EO @ 750 µg/kg, oral, once a week for 24 weeks + ProImmu @ 500 mg/kg, oral, daily for 12 weeks after 12 weeks of EO administration	25th	11.8a±0.0	5.70a±0.03

* Number of animals in each group=6.

Mean with same superscript does not differ significantly.

The corroborating findings in regard to this study have been reported by several investigators. The normal values (Group 1) of all the biochemical parameters estimated in the study correspond with the normal values cited by Madhuri (2008). The SGOT activity has been found to be elevated during cancer pathogenesis (Bradley *et al.*, 1972; Madhuri, 2008; Madhuri *et al.*, 2010; Shar and Kew, 1982; Somkuwar, 2003). The increased SLDH activity has been observed by Singh *et al.* (1973) after administration of oestrogen in rat. Similarly, Savlov *et al.* (1974) recorded higher levels of SLDH and G6PDH in women with ERs involved mammary carcinoma and benign breast tumour. Cornelis *et al.* (1997) noticed that G6PDH is strongly affected by contaminants and subsequent hepatocellular carcinogenesis, resulting into its reduced level in the blood. Somkuwar (2003) in mice bearing S180 cancer noted a significant (P<0.01) decline in the STP level and such a decline might be due to the fact that cancer had injured the liver cells, causing depletion of protein synthesis.

The significant or insignificant normalization in the activities of estimated LFTs caused by ProImmu is in close relation with the report of Das *et al.* (2000) who noticed the improvements by ProImmu (Immu-21) in different LFTs (*e.g.*, SGOT and SGPT). The significant beneficial effects in some LFTs have also been reported after treatment with some plant-ingredients of ProImmu, *viz.*, *O. sanctum* and *W. somnifera* against S180 cancer (Somkuwar, 2003), and *E. officinalis* in during liver damage (Sultana *et al.*, 2005). The results of this study suggest that the values of SGOT, SLDH, G6PDH and STP are significantly altered after administration of EO. However, the altered values are significantly returned to normal levels after administration of ProImmu, proving that this herbal drug has beneficial effects over the uterine and ovarian cancer (Madhuri, 2008; Madhuri *et al.*, 2010; Pandey *et al.*, 2013a).

Normalcy in Hematological Profile by Herbal Drugs in Estrogen Caused Cytotoxicity

Estrogen has been stated to cause many detrimental effects both in humans and animals. Excessive doses of estrogen may cause nausea, vomiting, anorexia, migraine, blurring of vision, mental depression, headache, asthma, endometriosis, fibroids, breast engorgement, increased vaginal secretion, oedema, cardiovascular and hepatic diseases, cancer, stroke, Alzheimer's disease, and many others in human beings (Loose and Stancel, 2006; Madhuri, 2008; Madhuri *et al.*, 2007a and 2012c; Pandey *et al.*, 2013a). The adverse effects such as bone marrow suppression, pyometra and infertility in bitches have been noticed after treatment with many estrogen preparations (Cain, 2001). Furthermore, estrogen has been found to cause lethargy, anorexia and collapse with clinical findings including vaginal discharge, endometritis and enlarged mammary glands in bitches (Acke *et al.*, 2003). The EO (a semisynthetic 17β-oestradiol being highly potent estrogen) in doses of 250, 500 and 750 μg per kg, orally, weekly has been reported to produce uterine and ovarian cytotoxicity and cancer in albino rats (Madhuri, 2008 and 2011; Madhuri *et al.*, 2012c).

Several ethnomedicinal plants and their preparations are being used for prevention and treatment of cytotoxicity caused by various toxicants. According to WHO estimates, more than 80 per cent people depend on the autochthonous herbal drugs for their primary health needs (Sivalokanathan *et al.*, 2005). ProImmu has been reported to possess immunomodulatory effect and restore the normal histoarchitecture of damaged tissues (Agrawala *et al.*, 2001; Madhuri, 2008). Cytogenic effect of ProImmu has been observed against EO induced uterine damage in rat (Madhuri and Pandey, 2007).

In view of above facts, a study was carried out (Madhuri, 2011; Madhuri *et al.*, 2012c) to evaluate the efficacy of three herbal drugs, *viz.*, *T. cordifolia*, *W. somnifera* and ProImmu against EO induced hematological alteration in female albino rats.

Experimental Study

Fifty-four healthy inbred female albino rats (each weighing 100-150 g) were used in the study. All the rats were quarantined for one week period after arrival. Thereafter, the rats were equally divided into nine groups (each group consisted of 6 rats). The animals were kept in cages under standard laboratory conditions in the animal house and fed on standard pellet diet and clean drinking water. However, the animals were fasted overnight before the experiment, but water was given *ad libitum*. The experimental designs and protocols in the study received the approval of the IAEC.

The lynoral tablets (containing 0.05 mg of EO only in each tablet) were triturated and suspended in distilled water mixed with a pinch of *Gum acacia* powder (since the drug is insoluble in water). The normal saline administered to the rats was also mixed with a pinch of *Gum acacia* powder to have uniformity with EO suspension. The hydroalcoholic extract (HAE) of the powdered *T. cordifolia* stem barks and *W. somnifera* roots was prepared separately with equal part of distilled water and ethyl alcohol (ethanol). Similarly, the aqueous suspension of ProImmu (mixed with a pinch of *Gum acacia* powder, since ProImmu is insoluble in water) was prepared prior to its administration to rats. Other required chemicals and reagents were purchased from the chemical shops.

The rats of group 1 were administered with normal saline to serve as normal group. EO (250 µg/kg, oral, thrice a week) was administered to the rats of groups 2 to 5 and groups 6 to 9 for 8 and 12 weeks, respectively. However, the HAE of *T. cordifolia* (250 mg/kg, oral, daily), *W. somnifera* (250 mg/kg, oral, daily) and aqueous suspension of ProImmu powder (150 mg/kg, oral, daily) were administered to the rats of groups 3 to 5, respectively for 8 weeks. Similarly, these three herbal drugs at the same dose rate were administered to the rats of groups 7 to 9, respectively for 12 weeks.

After the end of experiment, the hematological study in the blood samples of all the rats was performed. The blood was collected from the eye veins (intraocular method) of rats. The hematological parameters, *viz.*, Hb, TLC and differential leucocyte count (DLC, *viz.*, lymphocyte, monocyte, neutrophil, eosinophil and basophil) were estimated according to the methods described by Jain (1986). For statistical analysis, the data of each parameter were analyzed and the significance of difference was determined by using Duncan's new multiple range test at P=0.05 (5 per cent level of significance) in completely random design (Steel and Torrie, 1980).

The Hb concentration (g/dl), TLC (10^3/µl) and lymphocyte count (per cent) observed in the rats of groups 1 to 9 are presented in Table 15. The Hb values of groups 1, 5, 9, 4, 3, 8 and 7 did not decrease significantly (P<0.05) and found non-significant with each other. However, the Hb values of these groups decreased significantly with groups 2 and 6. The Hb values of groups 2 and 6 also decreased significantly with each other. There was found no significant difference amongst the TLC values of groups 1, 5, 9 and 4; groups 5, 9, 4, 3 and 8; groups 9, 4, 3, 8 and 7; groups 2 and 6. However, the TLC value differed significantly (P<0.05) in the remaining groups, *viz.*, group 1 with group 3, 8, 7, 2 and 6; group 5 with groups 7, 2 and 6; groups 9, 4, 3, 8 and 7 with groups 2 and 6. The lymphocyte values of groups 1, 5 and 9 did not decrease significantly (P<0.05) and found non-significant with each other. Similarly, the lymphocyte value of groups 5, 9 and 4 did not decrease significantly with each other; whilst such values of groups 9, 4 and 3; groups 4, 3, 8 and 7; groups 8, 7 and 2; groups 2 and 6 did not decrease significantly with each other. However, the lymphocyte values of group 1 with groups 4, 3, 8, 7, 2 and 6; group 5 with groups 3, 8, 7, 2 and 6; group 9 with groups 8, 7, 2 and 6; groups 4 and 3 with groups 2 and 6; groups 8 and 7 with group 6 differed significantly (P<0.05).

The concentrations (per cent) of monocyte, neutrophil and eosinophil estimated in the rats of groups 1 to 9 are depicted in Table 16. The monocyte value of group 1 differed significantly (P<0.05) with those of groups 2 to 9. Amongst the latter groups, the monocyte values differed significantly as under: group 5 with groups 3, 8, 7, 2 and 6; group 9 with groups 8, 7, 2, 6; group 4 with groups 7, 2, and 6; groups 3 and 8 with groups 2 and 6; groups 2 and 6 with all other groups except group 7. However, there was no significant difference amongst the remaining groups. The neutrophil values of groups 1 and 5 decreased significantly (P<0.05) in all other groups. Similarly, such values of groups 9 and 4 differed significantly with groups 3, 8, 7, 2 and 6; while the neutrophil values of groups 3, 8 and 7 differed significantly with groups 2 and 6. There was also found a significant difference between the neutrophil values of groups 2 and 6. The neutrophil values of groups 1 and 5; groups 9 and 4; groups 3, 8 and 7 did not differ significantly with each other. The eosinophil values of group 1 increased significantly (P<0.05) in groups 8, 7, 2 and 6. Such value of group 6 decreased significantly in all other groups. There was found no significant difference amongst the eosinophil values of groups 1, 5, 9, 4 and 3; groups 5, 9, 4, 3 and 8; groups 4, 3, 8 and 7; groups 8, 7 and 2. The basophil (per cent) in normal rats was found to be zero. Therefore, basophil in treated groups (2-9) was not counted.

The hematological study conducted in this work showed that EO altered levels of many parameters could be significantly normalized by ProImmu (150 mg/kg, orally, daily for 8 weeks); however, the normalization of these parameters brought up by *T. cordifolia* (250 mg/kg, oral, daily for 8 weeks) and *W. somnifera* (250 mg/kg, oral, daily for 8 weeks) was found to be of lesser degree. Further, ProImmu (150 mg/kg, oral, daily for 12 weeks), *T. cordifolia* (250 mg/kg, oral, daily for 12 weeks) and *W. somnifera* (250 mg/kg, oral, daily for 12 weeks) also caused the normalcy of these parameters, but to less extent than observed on the 8 weeks. These findings are in accordance with many authors. It can be, therefore, concluded that ProImmu, *T. cordifolia* and *W. somnifera* possess cytogenic effect, which is time dependent. The cytogenic effects of these herbal drugs may be due to their potent immunostimulatory

Table 15: Effect of *T. cordifolia*, *W. somnifera* and ProImmu on Hb, TLC and lymphocyte during EO induced cytotoxicity in rat

Group*	Treatment	Week of Experiment	Value (Mean#±SE)		
			Hb (g/dl)	TLC (10³/µl)	Lymphocyte (per cent)
1	Normal saline	1st	13.7ª±1.4	7766.7ª±647.9	36.8ª±1.0
2	EO @ 250 µg/kg, oral, thrice a week for 8 weeks	9th	10.0ᵇ±0.3	3800.0ᵈ±57.7	26.3ᵉᶠ±0.9
3	EO as per group 2 + *T. cordifolia* @ 250 mg/kg, oral, daily for 8 weeks from the start of experiment	9th	12.5ª±0.2	5866.7ᵇᶜ±566.1	31.5ᶜᵈ±0.8
4	EO as per group 2 + *W. somnifera* @ 250 mg/kg, oral, daily for 8 weeks from the start of experiment	9th	12.6ª±0.2	6316.7ᵃᵇᶜ±6892	33.0ᵇᶜᵈ±0.82
5	EO as per group 2 + ProImmu @ 150 mg/kg, oral, daily for 8 weeks from the start of experiment	9th	13.0ª±0.3	7300.0ᵃᵇ±587.7	35.5ᵃᵇ±1.1
6	EO @ 250 µg/kg, oral, thrice a week for 12 weeks	13th	7.8ᶜ±0.3	3150.0ᵈ±76.4	20.0±0.6
7	EO as per group 6 + *T. cordifolia* @ 250 mg/kg, oral, daily for 12 weeks from the start of experiment	13th	12.0ª±0.2	5300.0ᶜ±258.2	29.5ᵈᵉ±1.2
8	EO as per group 6 + *W. somnifera* @ 250 mg/kg, oral, daily for 12 weeks from the start of experiment	13th	12.3ª±0.2	5716.7ᵇᶜ±463.6	30.0ᵈᵉ±0.9
9	EO as per group 6 + ProImmu @ 150 mg/kg, oral, daily for 12 weeks from the start of experiment	13th	12.8ª±0.3	6483.3ᵃᵇᶜ±489.9	34.5ᵃᵇᶜ±0.8

* Number of animals in each group=6.

Mean with same superscript does not differ significantly.

Table 16: Effect of *T. cordifolia*, *W. somnifera* and Prolmmu on Monocyte, Neutrophil and Eosinophil during EO Induced Cytotoxicity in Rat

Group*	Treatment	Week of Experiment	Value (Mean#±SE)		
			Monocyte (per cent)	Neutrophil (per cent)	Eosinophil (per cent)
1	Normal saline	1st	7.0±0.4	63.3^a±2.5	4.1^e±0.5
2	EO @ 250 µg/kg, oral, thrice a week for 8 weeks	9th	1.5±0.2	37.0^f±0.5	6.5^b±0.4
3	EO as per group 2 + *T. cordifolia* @ 250 mg/kg, oral, daily for 8 weeks from the start of experiment	9th	3.5^{cde}±0.4	49.2^c±0.8	5.0^{cde}±0.6
4	EO as per group 2 + *W. somnifera* @ 250 mg/kg, oral, daily for 8 weeks from the start of experiment	9th	4.2^{bcd}±0.6	54.8^b±2.5	4.8^{cde}±0.3
5	EO as per group 2 + Prolmmu @ 150 mg/kg, oral, daily for 8 weeks from the start of experiment	9th	5.5^b±0.4	61.0^a±1.2	4.3^{de}±0.3
6	EO @ 250 µg/kg, oral, thrice a week for 12 weeks	13th	1.3±0.2	31.2^a±1.1	8.0±0.4
7	EO as per group 6 + *T. cordifolia* @ 250 mg/kg, oral, daily for 12 weeks from the start of experiment	13th	2.3^{ef}±0.2	45.3^c±1.7	6.0^{bc}±0.4
8	EO as per group 6 + *W. somnifera* @ 250 mg/kg, oral, daily for 12 weeks from the start of experiment	13th	3.3^{de}±0.5	47.0^c±1.7	5.5^{bcd}±0.2
9	EO as per group 6 + Prolmmu @ 150 mg/kg, oral, daily for 12 weeks from the start of experiment	13th	4.7^{bc}±0.4	56.8^b±1.9	4.5^{de}±0.4

* Number of animals in each group=6.

Mean with same superscript does not differ significantly.

and antioxidant, phagocytic and other cytoprotective activities (Madhuri, 2011; Madhuri *et al.*, 2012c; Pandey *et al.*, 2013a).

Normalization of Enzymes by Plant-Drugs in Ethinyl Oestradiol Caused Cytotoxicity

Ethinyl oestradiol (EO, an oestrogen) has been reported to cause severe cytotoxicity, including cancer both in humans and animals. Many plant-drugs have been used for prevention and treatment of cytotoxicity caused by various toxicants. For example, the plant-drugs, *viz.*, ProImmu, *Tinospora cordifolia* and *Withania somnifera* have been reported to elicit the cytogenic or anticancer effects, including the normalcy in enzymic activities and other LFTs altered during EO induced cytotoxicity/cancer (Madhuri, 2008 and 2011; Pandey *et al.*, 2013a).

Considering the above facts in view, a study was conducted (Madhuri, 2011) to evaluate the beneficial effects of *T. cordifolia*, *W. somnifera* and ProImmu on enzymic alterations (biochemical changes) during EO induced cytotoxicity in female albino rats.

Experimental Study

The experiment was conducted similar to the experimental designs and protocols mentioned under '*Normalcy in hematological profile by herbal drugs in estrogen caused cytotoxicity*' (above experiment). For biochemical study, the blood of all the rats was collected from their eye veins and the serum was separated. The enzymic activities of SGOT, (Bradley *et al.*, 1972), SGPT (Bradley *et al.*, 1972), SAP (Wilkinson *et al.*, 1969) and SLDH (Bergmeyer, 1971) were estimated on computerized serum auto-analyzer. The biochemical data of each parameter (LFT) were analyzed statistically, and the mean values and standard error (SE) were calculated. To find out the significance of difference among different groups of treatments, Duncan's new multiple range test at P=0.05 (5 per cent level of significance) in completely random design was used (Steel and Torrie, 1980).

The results of all the biochemical parameters estimated in the study are summarized in Table 17. The SGOT, SGPT, SAP and SLDH activities (IU/L) of normal rats (Group 1) observed on the 1st week significantly (P<0.05) increased in group 2 on the 9th week and in group 6 on the 13th week after EO administration. The increased activities of all these parameters of groups 2 and 6 significantly decreased in groups 3 to 5 and groups 7 to 9, respectively. Among groups 3 to 5, the SGOT, SGPT and SAP values of groups 3 and 5; and groups 4 and 5 differed significantly. However, the SLDH values of groups 3 to 5 differed significantly with each other. As regard to increased enzymic activities of all parameters of group 6 are concerned, these significantly decreased in groups 7 to 9. Among groups 7 to 9, the SGOT and SGPT values of groups 7 and 9; and groups 8 and 9 differed significantly. However, the SAP and SLDH values of groups 7 to 9 differed significantly with each other. There was found no significant difference between the SGPT, SAP and SLDH activities of groups 1 and 5. Similarly, there was no significant difference between the SGOT, SGPT and SAP activities of groups 3 and 4. Likewise, no significant difference was found between the SGOT, SGPT and SLDH activities of groups 3 and 8. Furthermore,

Table 17: Effect of *T. cordifolia*, *W. somnifera* and Prolmmu on SGOT, SGPT, SAP and SLDH Enzymes during EO Induced Cytotoxicity in Rat

Group*	Treatment	Week of Experiment	Mean#±SE value (IU/L)			
			SGOT	SGPT	SAP	SLDH
1	Normal saline	1st	$72.9^a\pm1.4$	$34.0^a\pm0.1$	$164.7^a\pm1.8$	$167.2^a\pm1.4$
2	EO @ 250 µg/kg, oral, thrice a week for 8 weeks	9th	$118.1^b\pm1.7$	$72.5^b\pm2.2$	$230.3^b\pm0.7$	$230.7^b\pm0.9$
3	EO as per group 2 + *T. cordifolia* @ 250 mg/kg, oral, daily for 8 weeks from the start of experiment	9th	$103.8^{de}\pm1.0$	$54.7^{de}\pm2.3$	$185.0^e\pm1.4$	$190.3^d\pm1.3$
4	EO as per group 2 + *W. somnifera* @ 250 mg/kg, oral, daily for 8 weeks from the start of experiment	9th	$98.7^e\pm3.9$	$50.3^e\pm2.6$	$182.7^e\pm0.9$	$183.8^e\pm1.0$
5	EO as per group 2 + Prolmmu @ 150 mg/kg, oral, daily for 8 weeks from the start of experiment	9th	$88.8^f\pm1.2$	$38.8^{fg}\pm2.6$	$167.8^g\pm2.2$	$171.0^g\pm1.0$
6	EO @ 250 µg/kg, oral, thrice a week for 12 weeks	13th	$125.5^a\pm1.4$	$79.5^a\pm0.8$	$249.7^a\pm1.5$	$247.5^a\pm1.1$
7	EO as per group 6 + *T. cordifolia* @ 250 mg/kg, oral, daily for 12 weeks from the start of experiment	13th	$111.7^c\pm1.7$	$65.3^c\pm1.4$	$200.5^c\pm2.0$	$201.3^c\pm1.7$
8	EO as per group 6 + *W. somnifera* @ 250 mg/kg, oral, daily for 12 weeks from the start of experiment	13th	$108.0^{cd}\pm1.7$	$59.5^{cd}\pm1.8$	$192.0^d\pm1.0$	$195.2^d\pm1.5$
9	EO as per group 6 + Prolmmu @ 150 mg/kg, oral, daily for 12 weeks from the start of experiment	13th	93.1 ± 1.1	44.0 ± 1.0	$174.3^f\pm1.5$	$178.5^f\pm3.0$

* Number of animals in each group=6.

Mean with same superscript does not differ significantly.

there was observed no significant difference between the SGPT activities of groups 5 and 9; and between the SGOT and SGPT activities of groups 7 and 8.

The normal values of SGOT, SGPT and SAP recorded in this research correspond with the normal values reported by Bhalerao (2006) and Madhuri (2008) in rat, and Pandey (1990) in rabbit. Similarly, the normal value of SLDH is comparable with those reported by Bergmeyer (1971) in humans and animals, and Madhuri (2008) in rat. The increased activities of SGOT, SGPT and SAP, as noticed in this study, may be correlated with the reports of Benjamin and McKelvie (1978), Bhalerao (2006) and Pandey (1990), who explained the increased activities of both these enzymes on exposure to hepatotoxins, including drugs and chemicals. Shar and Kew (1982) have reported the elevated activities of these enzymes during OCs (containing oestrogen) induced cancer in woman. Benjamin and McKelvie (1978) and Bergmeyer (1971) cited that SLDH is elevated in humans and animals during liver damage. As observed in this study, Madhuri (2008) and Singh *et al.* (1973) also noticed the increased SAP and SLDH activities after administration of oestrogen or EO in rat.

The results of the study indicated that SGOT, SGPT, SAP and SLDH activities after treatment with *T. cordifolia*, *W. somnifera* and ProImmu significantly decreased against EO induced cytotoxicity and returned towards normal level, especially after treatment with ProImmu. Further, the SGPT SAP and SLDH activities of group 5 exactly returned to normal, as observed in group 1 (normal). The findings of the study are in accordance with many investigators. The improvement in SGOT and SGPT activities caused by ProImmu was also seen by Das *et al.* (2000). Similar results of SGPT and SLDH were observed by Sultana *et al.* (2005) who reported significant decrease in carbon tetrachloride (a hepatotoxicant) increased SGPT and SLDH activities of rat pretreated with *E. officinalis* (an ingredient of ProImmu). The improvement in SAP and SLDH activities brought up by ProImmu was observed by Madhuri (2008). The decrease in SAP activity leading to normalization was also observed by Somkuwar (2003) after administration of *O. sanctum* and *W. somnifera* (both plants are ingredients of ProImmu) in mice bearing S180 tumour. The cytogenic effects (including normalization of SGOT, SGPT, SAP and SLDH activities) of ProImmu have been observed against EO induced cytotoxicity, leading to cancer in rats (Madhuri, 2008; Madhuri *et al.*, 2012f).

Tissue Repair of Uterus and Ovary by Plant-Drugs in Oestrogen Produced Cytotoxicity

A study (Madhuri, 2011) has been done to assess the cytogenic effect of certain plant-drugs (*viz.*, *T. cordifolia*, *W. somnifera* and ProImmu) against oestrogen (EO) produced cytotoxicity in the uterus and ovary of albino rat. The cytogenic effect of plant-drugs and cytotoxic effect of EO have been evaluated by conducting histopathological study of uterine and ovarian tissues.

Experimental Study

The experiment was done similar to the experimental designs and protocols mentioned under *'Normalcy in hematological profile by herbal drugs in estrogen caused cytotoxicity'*. For histopathological study, the rats were sacrificed by cervical dislocation (euthanized scientifically). The uteri and ovaries of all the rats/groups were collected and preserved in 10 per cent buffered formalin. The uterine and ovarian tissues were processed and stained with H&E stain as per the method of Culling (1963), and then the tissues were examined, microscopically for cytotoxic as well as cytogenic changes or normal histological profiles.

On the 1st week of experiment, the microscopic examination of uterus (Figure 23) and ovary (Figure 26) of rats administered with normal saline (Group 1) have shown normal histological profiles.

On the 9th week of experiment, the uterine tissues of the rats of group 2 administered with EO (250 µg/kg, oral, thrice a week for 8 weeks) have revealed congestion, proliferation of fibrovascular connective tissues, and necrosis and desquamation of glandular epithelial cells (Figure 24). At the same period, the ovarian tissues of group 2 have shown congestion and infiltration of lymphocytes in interstitial tissues (Figure 27). However, *T. cordifolia* (250 mg/kg, oral, daily for 8 weeks) in the uterus and ovary of group 3, *W. somnifera* (250 mg/kg, oral, daily for 8 weeks) in the uterus and ovary of group 4 and ProImmu (150 mg/kg, oral, daily for 8 weeks) in the uterus (Figure 25) and ovary (Figure 28) of group 5 have produced regeneration and normalization in increasing order at many places. Besides, the lesser to least degree of damage has been noticed as compared to that observed in the uterine and ovarian tissues of group 2.

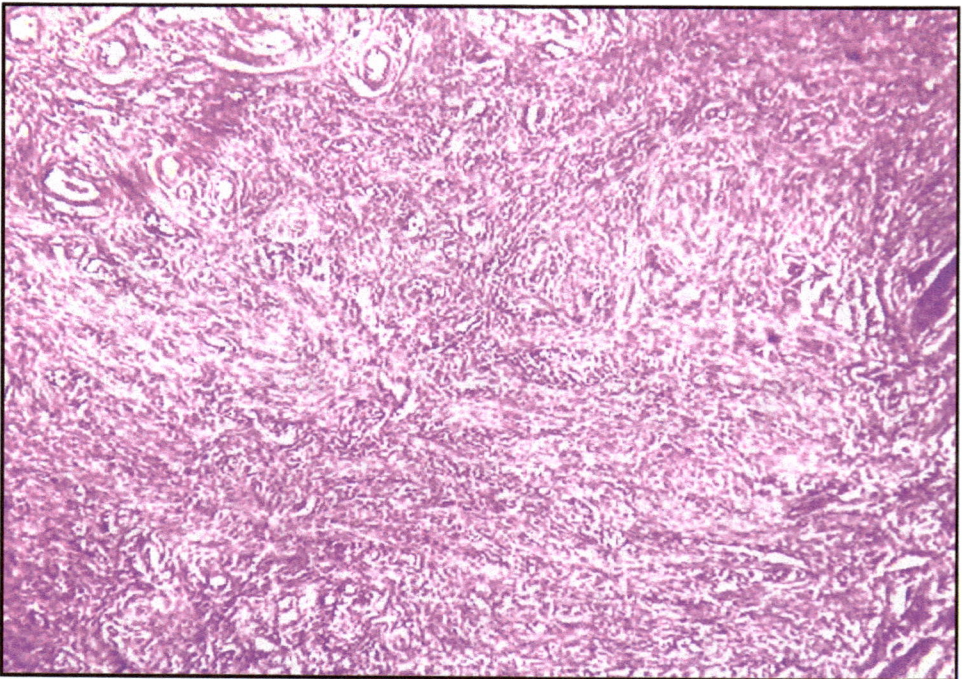

Figure 23: Uterus of Rat, Showing Normal Histological Profile (x100, H&E) [Madhuri, 2011].

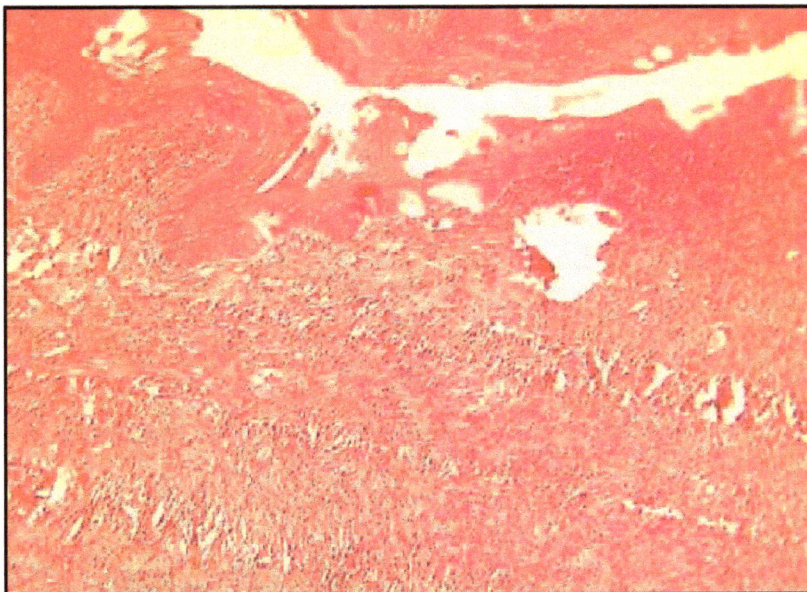

Figure 24: Uterus of Rat on 9th Week of EO (250 µg/kg, oral, thrice a week for 8 weeks) Administration, Showing Congestion, Proliferation of Fibrovascular Connective Tissues, Necrosis and Desquamation of Glandular Epithelial Cells (x100, H&E) [Madhuri, 2011].

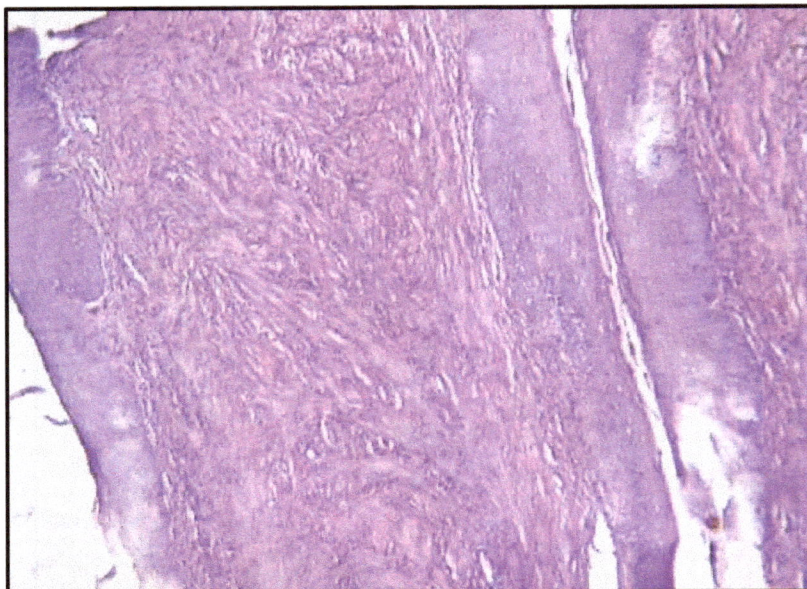

Figure 25: Uterus of Rat (Group 5) on 9th Week of EO (250 µg/kg, oral, thrice a week for 8 weeks) and Prolmmu (150 mg/kg, oral, daily for 8 weeks) Administration, Showing Very Mild Histopathological Changes and most of the Tissues are Repaired and Normalized (x100, H&E) [Madhuri, 2011].

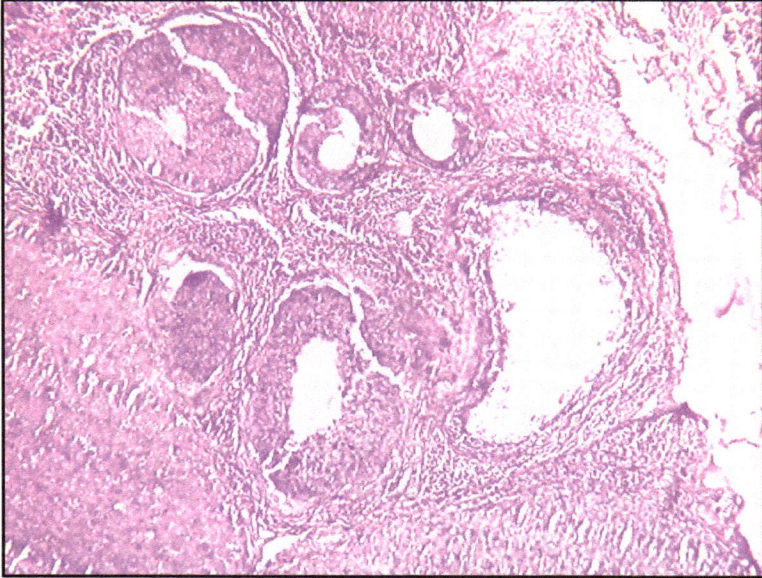

Figure 26: Ovary of Rat, Showing Normal Histological Profile (x100, H&E) [Madhuri, 2011].

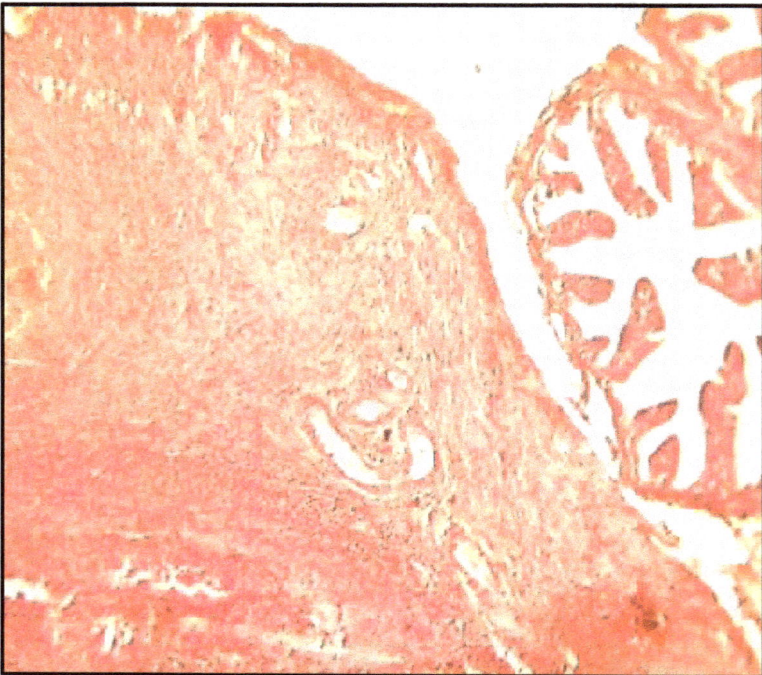

Figure 27: Ovary of Rat (Group 2) on 9th Week of EO (250 µg/kg, oral, thrice a week for 8 weeks) Administration, Showing Congestion and Infiltration of Lymphocytes in Interstitial Tissues (x100, H&E) [Madhuri, 2011].

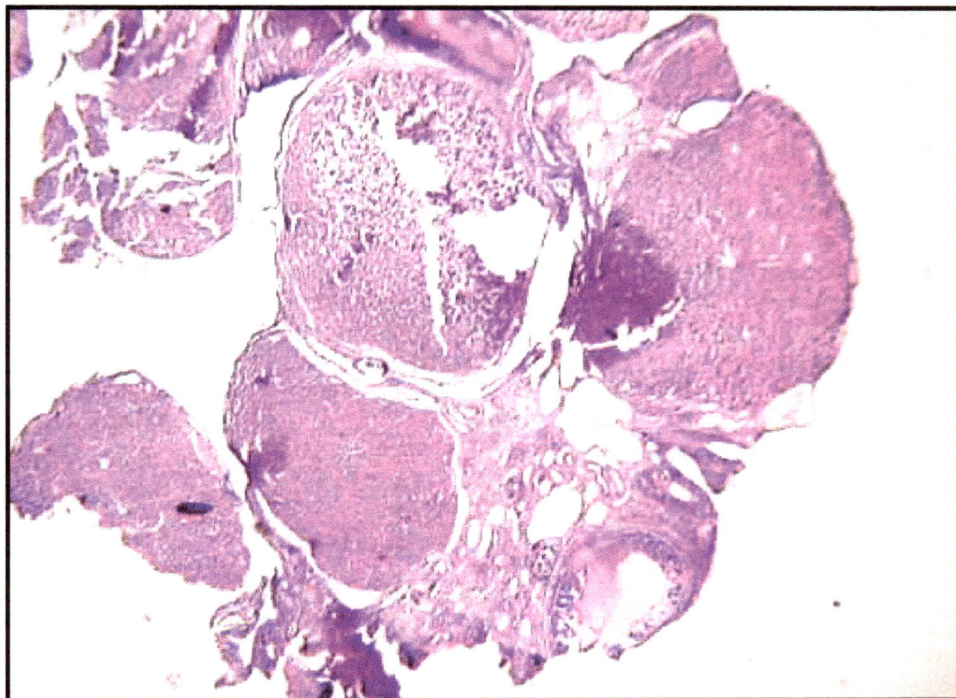

Figure 28: Ovary of Rat (Group 5) on 9ᵗʰ Week of EO (250 µg/kg, oral, thrice a week for 8 weeks) and ProImmu (150 mg/kg, oral, daily for 8 weeks) Administration, Showing Lesser Degree of Histopathological Changes and Regeneration and Normalization of Tissues are at many Places (x100, H&E) [Madhuri, 2011].

On the 13ᵗʰ week of experiment, the uterine tissues of group 6 dosed with EO (250 µg/kg, oral, thrice a week for 12 weeks) have shown similar nature of histopathological changes as observed in group 2, but the changes have been seen more severe and extensive. The fibrosis has resulted into compression of the endometrial glands. At the same period, the ovarian tissues of group 6 have revealed more prominent and severe changes as compared to all other groups (*i.e.,* groups 1-5 and 7-9). The vascular congestion and fibrosis in ovarian tissues have been observed at many places. The ovarian parenchyma has been replaced by fibrovascular connective tissues. The swelling of endothelial cells, thickening of blood vessel walls, extensive infiltration of lymphocytes, degeneration of follicular tissues and proliferation of fibrous tissues have also ppeared. However, the uteri and ovaries of groups 3, 4 and 5 treated with EO (250 µg/kg, oral, thrice a week for 12 weeks) concomitant with *T. cordifolia* (250 mg/kg, oral, daily for 12 weeks), *W. somnifera* (250 mg/kg, oral, daily for 12 weeks) and ProImmu (150 mg/kg, oral, daily for 12 weeks), respectively have shown regeneration and normalization in increasing order at many places; however, the less intensity of cytotoxic changes has occurred when compared with those observed in group 6.

The uterine and ovarian cytotoxicity produced by EO are in close conformity with the uterine and ovarian cytotoxicity as reported by many workers. Hertz (1976), Liehr (2001), Loose and Stancel (2006), Madhuri (2008 and 2011) and Satoskar *et al.* (2005) clearly pointed out that HRT, OCs or different oestrogen preparations may cause uterine and ovarian cytotoxicity/cancer in humans and different species of animals. Jabara (1962) reported that the endometrium of bitches after prolonged administration of stilbestrol oestrogen showed degeneration with marked glandular atrophy. Kader *et al.* (1969) reported that the thickened endometrium with numerous glands developed in animals receiving contraceptives for short periods; while atrophic endometrium or squamous metaplasia of endometrium with some cases of cystic dilatation of the glands produced after a long period of contraceptive therapy. Schwartz *et al.* (1969) observed that high and prolonged dose of quinestrol oestrogen caused uterine enlargement with endometrial hyperplasia and myometrial hypertrophy in bitches. Fayrer-Hosken *et al.* (1992) observed follicular ovarian cysts, and cystic endometrial hyperplasia and pyometra in bitch with hyperoestrogenism. Endometritis in bitches treated with oestrogen was reported by Acke *et al.* (2003). Further, similar to the present study, EO caused severe liver cytotoxicity (*i.e.*, hepatotoxicity) in female albino rats (Pandey *et al.*, 2008).

The results regarding tissue repairment produced by ProImmu, *T. cordifolia* and *W. somnifera* against EO induced uterine and ovarian cytotoxicity may be correlated with the reports of several workers. ProImmu has been reported to possess better immunocompetence since it enhances the activity of cytotoxic T cells as well as NK cells. It potentiates the specific and nonspecific immunity of the host. ProImmu increases the size, number and phagocytic activity of macrophages (*e.g.*, lymphocytes, monocytes, etc.). It also increases the lipopolysaccharide induced leucocyte (*e.g.*, lymphocytes and monocytes) proliferation. The stimulation of splenocytes to produce plaque forming cells by ProImmu helps in stimulating humoral arm of immunity. Due to all these activities, ProImmu may have cytogenic and tumoricidal effects (Agrawala *et al.*, 2001; Nemmani *et al.*, 2002). The cytogenic effects of ProImmu have been observed against EO induced cytotoxicity, leading to cancer in rats (Madhuri, 2008; Madhuri *et al.*, 2012b). ProImmu has been, therefore, reported to exhibit immunostimulatory, phagocytic, antioxidant and various other tissue protective activities (Madhuri, 2008). The cytogenic effect of ProImmu may also be due to its plant-ingredients, which have been reported as potent immunostimulatory and antioxidant plants (Chopra *et al.*, 2002; Pandey and Madhuri, 2006a; Prajapati *et al.*, 2003). *E. officinalis* fruit (Amla) has elicited immunostimulatory and antitumour effects (Jeena *et al.*, 2001). The methanolic extract (200 mg/kg, ip, daily for 5 days) of *T. cordifolia* stem has increased humoral immune response and reduced the solid tumour growth in mice (Mathew and Kuttan, 1999). The anticancer/cytogenic effect of *T. cordifolia* has also been reported by Singh *et al.* (2005). Similar effects of *W. somnifera* have been observed by many workers, including Gupta *et al.* (2001), Madhuri and Pandey (2009c), and Somkuwar (2003). The immunomodulatory effect of *W. somnifera* and *T. cordifolia* has been noticed by Thatte and Dahanukar (1989).

It can be, therefore, concluded that the herbal drugs ProImmu, *T. cordifolia* and *W. somnifera* possess cytogenic effect, which is time dependent. The cytogenic effect and mechanism of action of ProImmu, *T. cordifolia* and *W. somnifera* may be postulated on the basis of their immunostimulatory, antioxidant, phagocytic and other cytoprotective activities (Madhuri, 2011).

Chapter 8

Images of Important
Anticancer Medicinal Plants

Although there are more than 1,000 plants that possess significant anticancer activity; however, the images (photographs/figures) of only 88 important anticancer medicinal plants (Figures 29-116) have been presented here. These figures have been extracted from different internet websites (particularly *'google.com'*) that are duly acknowledged.

Figure 29: *Abelmoschus esculentus*

Figure 30: *Abrus precatorius*

Figure 31: *Agave Americana*

Figure 32: *Aglaia elaeagnoidea*

Figure 33: *Allium cepa*

Figure 34: *Allium sativum*

Figure 35: *Aloe vera*

Figure 36: *Andrographis paniculata*

Figure 37: *Azadirachta indica*

Figure 38: *Berberis aristata*

Figure 39: *Brassica campestris*

Figure 40: *Brassica oleracea* var. *botrytis*

Figure 41: *Brassica oleracea* var. *capitata*

Figure 42: *Camellia sinensis*

Figure 43: *Camptotheca acuminata*

Figure 44: *Cannabis sativa*

Figure 45: *Cassia absus*

Figure 46: *Cassia fistula*

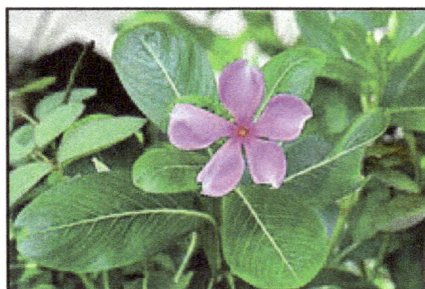

Figure 47: *Catharanthus roseus*   Figure 48: *Citrus limon*

Figure 49: *Clerodendrum infortunatum*

Figure 50: *Coriandrum sativum*

Figure 51: *Crocus sativus*

Figure 52: *Curcuma longa*

Figure 53: *Cucumis sativus*

Figure 54: *Daucus carota*

Figure 55: *Emblica officinalis*

Figure 56: *Ervatamia heyneana*

Figure 57: *Erythronium americanum*

Figure 58: *Eugenia caryophyllata*

Figure 59: *Fragaria vesca*

Figure 60: *Ganoderma lucidum*

Figure 61: *Glycerrhiza glabra*

Figure 62: *Glycine javanica*

Figure 63: *Hagenia abyssinica*

Figure 64: *Glycine max*

Figure 65: *Hygrophila spinosa*

Figure 66: *Indigofera aspalathoides*

Figure 67: *Indigofera mysorensis*

Figure 68: *Jatropha gossypifolia*

Figure 69: *Kaempferia rotunda*

Figure 70: *Lantana camara*

Figure 71: *Lonicera japonica*

Figure 72: *Lycopersicon esculentum*

Figure 73: *Martynia annua*

Figure 74: *Malus domestica*

Figure 75: *Mentha arvensis*

Figure 76: *Momordica charantia*

Figure 77: *Morinda citrifolia*

Figure 78: *Moringa oleifera*

Figure 79: *Nelumbo nucifera*

Figure 80: *Ocimum sanctum*

Figure 81: *Olea europaea*

Figure 82: *Oldenlandia diffusa*

Figure 83: *Operculina turpethum*

Figure 84: *Panax ginseng*

Figure 85: *Patrinia heterophylla*

Figure 86: *Phaseolus vulgaris*

Figure 87: *Pinus pinaster*

Figure 88: *Piper longum*

Figure 89: *Plumbago rosea*

Figure 90: *Podophyllum hexandrum*

Figure 91: *Polygonatum multiflorum*

Figure 92: *Polygonum cuspidatum*

Figure 93: *Prunella vulgaris*

Figure 94: *Rheum palmatum*

Figure 95: *Rhus chinensis*

Figure 96: *Ricinus communis*

Figure 97: *Rubia cordifolia*

Figure 98: *Scutellaria radix*

Figure 99: *Semecarpus anacardium*

Figure 100: *Silybum marianum*

Figure 101: *Smilax chinensis*

Figure 102: *Solanum aculeastrum*

Figure 103: *Solanum dulcamara*

Figure 104: *Stereospermum personatum*

Figure 105: *Swertia chirata*

Figure 106: *Taxus brevifolia*

Figure 107: *Terminalia arjuna*

Figure 108: *Terminalia catappa*

Figure 109: *Tinospora cordifolia*

Figure 110: *Trigonella foenumgraecum*

Figure 111: *Vanda parviflora*

Figure 112: *Viscum album*

Figure 113: *Vitis rotundifolia*

Figure 114: *Vitex negundo*

Figure 115: *Withania somnifera*

Figure 116: *Zingiber officinale*

Bibliography

Acke, E., Mooney, C.T. and Jones, B.R. (2003). Oestrogen toxicity in a dog. Irish Vet. J., 56(9): 465-468.

Agarwal, C., Sharma, Y. and Agarwal, R. (2000). Anticarcinogenic effect of a polyphenolic fraction isolated in human prostrate carcinoma DU145 cells: Modulation of cell cycle regulators and induction of G_1 arrest. Mol. Carcinog, 28: 129-138.

Agarwal, R., Agarwal, C., Ichikawa, H., Singh, R.P. and Aggarwal, B.B. (2006).

Anticancer potential of silymarin: From bench to bed side. Anticancer Res., 26(6B): 4457-4498.

Agrawala, S.K., Chatterjee, S. and Misra, S.K. (2001). Immune-potentiation activity of a polyherbal formulation "Immu-21" (Research name). Phytomedica, 2(1 and 2): 1-22.

Ahilan, B., Nithiyapriyatharshini, A. and Ravaneshwaran, K. (2010). Influence of certain herbal additives on the growth, survival and disease resistance of goldfish, *Carassius auratus* (Linnaeus). Tamilnadu J. Vet. Ani. Sci., 6(1): 5-11.

Ahmad, N., Feyes, D.K., Nieminen, A.L., Agarwal, R. and Mukhtar, H. (1997). Green tea constituent epigallocatechin-3-gallet and induction of apoptosis and cell cycle arrest in human carcinoma cells. J. Natl. Cancer Inst., 89: 1881-1886.

Ainsworth, A.J., Dexiang, C. and Waterstrat, P.R. (1991). Changes in peripheral blood leukocyte percentages and function of neutrophils in stressed channel catfish. J. Aquatic Anim. Health, 3: 41-47.

American Cancer Society (2006). www.cancervax.com/info/index.htm.

Andallu, B. and Radhika, B. (2000). Hypoglycemic, diuretic and hypocholesterolemic effect of winter cherry (*Withania somnifera* Dunal) root. Indian J. Exp. Biol., 38: 607-609.

Anilakumar, K.R., Nagaraj, N.S. and Santhanam, K. (2007). Reduction of hexachlorocyclohexane-induced oxidative stress and cytotoxicity in rat liver by *Emblica officinalis* Gaertn. Indian J. Exp. Biol., 45(5): 450-454.

Arivazhagan, S., Velmurugan, B., Bhuvaneswari, V. and Nagini, S. (2004). Effect of aqueous extract of garlic (*Allium sativum*) and neem (*Azadirachta indica*) leaf containing blood oxidant-antioxidant status during experimental gastric carcinogenesis. J. Med. Food, 7(3): 334-339.

Arora, S., Dhillon, S., Rani, G. and Nagapal, A. (2004). The *in vitro* antibacterial/ synergistic activities of *Withania somnifera* extracts. Fitoterapia, 75: 385-388.

Aruna, K. and Sivaramkrishnan, V.M. (1990). Plant products as protective agents against cancer. Indian J. Exp. Biol., 28: 1008-1011.

Babbar, O.P., Bajpai, S.K., Choudhari, B.L. and Khan, S.K. (1979). Occurrence of infection like antiviral and antitumour factors(s) in extracts of some indigenous plants. Indian J. Exp. Biol., 17: 451-454.

Bai, W., Oliveros-Saunders, B., Wang, Q., Acevedo-Duncan, M.E. and Nicosia, S.V. (2000). Estrogen stimulation of ovarian surface epithelial cell proliferation. In vitro Cell Dev. Biol. Anim., 36: 657-666.

Balachandran, P. and Govindarajan, R. (2005). Cancer: An Ayurvedic perspective. Pharmacol. Res., 51(1): 19-30.

Banu, S.M., Selvendiran, K., Singh, J.P. and Sakthisekaran, D. (2004). Protective effect of *Emblica officinalis* ethanolic extract against 7,12-dimethylbenz(a)anthracene (DMBA) induced genotoxicity in Swiss albino mice. Hum. Exp. Toxicol., 23(11): 527-531.

Beniashvili, D., Avinoachm, I., Baasov, D. and Zusman, I. (2005). The role of household electromagnetic fields in the development of mammary tumors in women: Clinical case-record observations. Med. Sci. Monit., 11: 10-13.

Benjamin, M.M. and McKelvie, D.H. (1978). *Clinical Biochemistry*. In: *Pathology of Laboratory Animals*, Vol. II (Benirschke, K., Garner, F.M. and Jones, T.C., eds.). Springer-Verlag and New York Inc. pp. 1750-1815.

Bergmeyer, H.U. (1971). In: *Methods of Enzymatic Analysis*, 2nd Edn. (Reprint). Verlag Chemie, G.M., B.H., Academic Press, New York. pp. 651-712.

Bhalerao, N. (2006). Studies on antihepatoxic activity of *Boerhaavia diffusa* (Punarnava) on experimental liver damage in rat. MVSc and AH thesis. JNKVV, Jabalpur, MP, India.

Bhatia, A.L. (2008). Effect of radiation and gene mediated radiation protection with special reference to melatonin. Proceedings: 1-51 (Part II, Abstracts- Animal, Veterinary and Fishery Sciences). 95th Indian Science Congress of the Indian Science Congress Association held at Visakhapatnam, AP, India.

Bhatia, N., Zhao, J., Wolf, D.M. and Agarwal, R. (1999). Inhibition of human carcinoma cell growth and DNA synthesis by silibinin, an active constituent of milk thistle: Comparison with silymarin. Cancer Lett., 147: 77-84.

Bhattacharya, S.K., Sanyal, A.K. and Ghosal, S. (1976). *Drugs and Central Synaptic Transmitters* (Bradely, P.B. and Dhawan, B.N., eds.). Macmillan, New Delhi, India.

Bradley, D.W., Maynard, J.E., Emery, G. and Webster, H. (1972). Transaminase activities in serum of long-term hemodialysis patients. Clin. Chem., 18: 1442.

Bunton, T.E. (1996). Experimental chemical carcinogenesis in fish. Toxicol. Pathol., 24(5): 603-618.

Butel, J.S. (2000). Viral carcinogenesis: Revelation of molecular mechanisms and etiology of human disease. Carcinogenesis, 21(3): 405-426.

Button, B.J. and Patel, N. (2004). Phytoestrogens for osteoporosis. Clin. Rev. Bone Min. Met., 2(4): 341-356.

Cain, J.L. (2001). Rational use of reproductive hormones. In: *Small Animal Clinical Pharmacology and Therapeutics* (Boothe, D.M., ed.). Saunders, Philadelphia. pp. 677-690.

Chabner, B.A., Amrein, P.C., Druker, B.J., Michaelson, M.D., Mitsiades, C.S., Goss, P.E., Ryan, D.P., Ramachandra, S., Richardson, P.G., Supko, J.G. and Wilson, W.H. (2006). Antineoplastic agents. In: *Goodman and Gilman's The Pharmacological Basis of Therapeutics*, 11th Edn. (Brunton, L.L., ed.). McGraw-Hill Co., New York. pp. 1315-1403.

Chansue, N., Ponpornpisit, A., Endo, M., Sakai, M. and Satoshi, Y. (2000). Improved immunity of tilapia *Oreochromis niloticus* by C-UP III, a herb medicine. Fish Pathol., 35: 89-90.

Chapekar, M.S. and Sahasrabudhe, M.B. (1981). Mode of action of GCE: An active anticancer principle isolated from an indigenous plant *Gymnosporia rothiana* Laws. Indian J. Exp. Biol., 19: 333-336.

Chitnis, M.P., Khandipkar, D.D., Adwankar, M.K. and Sahastrabudhe, M.B. (1971). Antitumour activity of the extract of root, stem and leaf of *Ervatamia heyneana*. Indian J. Exp. Biol., 9: 268-270.

Chopra, R.N., Nayar, S.L. and Chopra, I.C. (2002). *Glossary of Indian Medicinal Plants*. Council of Scientific and Industrial Research, New Delhi, India.

Christina, A.J., Joseph, D.G., Packialakshmi, M., Kothai, R., Robert, S.J.,

Chidambaranathan, N. and Ramasamy, M. (2004). Anticarcinogenic activity of *Withania somnifera* Dunal against Dalton's ascitic lymphoma. J. Ethnopharmacol., 93: 359-361.

Christybapita, D., Divyagnaneswari, M. and Michael, R.D. (2007). Oral administration of *Eclipta alba* leaf aqueous extract enhances the non-specific immune responses and disease resistance of *Oreochromis mossambicus*. Fish Shellfish Immunol., 23(4): 840-852.

Cleaver, J.E. and Crowley, E. (2002). UV damage, DNA repair and skin cancinogenesis. Front Biosci., 7: 1024-1043.

Cornelis, J.F., Van, N., Sieglinde, B. and Angela, K. (1997). Adaptational changes in kinetic parameters of G6PDH but not of PGDH during contamination-induced carcinogenesis in livers of North Sea flatfish. Biochimica et Biophysica Acta (Protein Structure and Mol. Enzymol.), 1342(2): 141-148.

Cougot, D., Buendia, M. and Neuveut, C. (2008). Carcinogenesis induced by hepatitis B virus. Human Cancer Viruses, 1: 108-136.

CSIR (1986). The Useful Plants of India. Publications and Information Directorate, Council of Scientific and Industrial Research, New Delhi, India.

Culling, C.F.A. (1963). Hand Book of Histological Techniques, 2nd Ed. Butterworth and Co. Ltd., London. pp. 25-172.

Cunat, S., Hoffmann, P. and Pujol, P. (2004). Estrogens and epithelial ovarian cancer. Gynecol. Oncol., 94(1): 25-32.

Cutler, B.S., Forbes, A.P., Ingersoll, F.M. and Scully, R.E. (1972). Endometrial carcinoma after stilbestrol therapy in gonadal dysgenesis. N. Engl. J. Med., 287: 628-631.

Das, B., Kundu, J., Bachar, S.C., Uddin, M.A. and Kundu, J.K. (2007). Anticancer and antibacterial activity of ethyl acetate extract of Ludwigia hyssopifolia Linn. and its active principle piperine. Pak. J. Pharm. Sci., 20(2): 128-131.

Das, S.N., Singh, J. and Agrawala, S.K. (2000). Chronic toxicity study of Immu-21. Phytomedica, 21: 89-94.

Das, S.K. and Vasudevan, D.M. (2006). Tulsi: The Indian holy power plant. Nat. Prod. Rad., 5: 279-283.

Davis, L. and Kuttan, G. (2002). Effect of Withania somnifera on cell mediated immune responses in mice. J. Exp. Clin. Cancer Res., 21: 585-590.

Dhar, M.L., Dhar, M.M., Dhawan, B.N., Mehrotra, B.N. and Roy, C. (1968). Screening of Indian plants for biological activity: Part I. Indian J. Exp. Biol., 6: 232-247.

Dikasso, D., Makonnen, E., Debella, A., Abebe, D., Urga, K., Makonnen, W., Melaku, D., Kassa, M. and Gupta, M. (2007). Antimalarial activity of Withania somnifera L. Dunal extract in mice. Pak. J. Pharm. Sci., 20: 231-235.

Dobaradaran, S., Naddafi, K., Nazmara, S. and Ghaedi, H. (2010). Heavy metals (Cd, Cu, Ni and Pb) content in two fish species of Persian Gulf in Bushehr Port. Iran. Afr. J. Biotechnol., 9(37): 6191-6193.

Dolcetti, R. and Masucci, M.G. (2003). Epstein-Barr virus: Induction and control of cell transformation. J. Cell Physiol., 196: 207-218.

Dong, Y., Yang, M.M. and Kwan, C.Y. (1997). In vitro inhibition of proliferation of HL-60 cells by tetrandrine and coriolus versicolor peptide derived from Chinese medicinal herbs. Life Sci., 60: 135-140.

Dugenci, S.K., Arda, N. and Candan, A. (2003). Some medicinal plants as immunostimulant for fish. J. Ethanopharmacol., 88(1): 99-106.

Duong Van Huyen, J.P., Sooryanarayana, V., Delignat, S., Bloch, M.F., Kazatchkine, M.D. and Kaveri, S.V. (2001). Variable sensitivity of lymphoblastoid cells to apoptosis induced by *Viscum album* Qu FrF, a therapeutic preparation of mistletoe lectin. Chemotherapy, 47: 366-376.

Eberhardt, M.V., Lee, C.Y. and Lui R.H. (2000). Antioxidant activity of fresh apples. Nature, 405: 903-904.

Edmondson, H.A., Henderson, B. and Benton, A. (1976). Liver cell adenomas associated with the use of oral contraceptives. N. Engl. J. Med., 294: 470-472.

Emami Khansari, F., Ghazi-Khansari, M. and Abdollahi, M. (2005). Heavy metals content of canned tuna fish. Food Chem., 93: 293-296.

Epstein, M.A. (2001). Reflections on Epstein-Barr virus: Some recently resolved old uncertainties. J. Infect., 43: 111-115.

Evans, W.C. (1989). *Trease and Evan's Pharmacognosy*, 13th Edn. Bailliere Tindall, London.

Fayrer-Hosken, R.A., Durham, D.H., Allen, S., Miller-Liebl, D.M. and Caudle, A.B. (1992). Follicular cystic ovaries and cystic endometrial hyperplasia in a bitch. J. Am. Vet. Med. Assoc., 201(1): 107-108.

Ferguson, L. and Denny, W. (1999). Carcinogen. In: *Encyclopedia of Molecular Biology*, Vol. I (Creighton, T.E., ed.). A Wiley Interscience Publication, New York. pp. 346-347.

Flack, C.P. and Woollen, J.W. (1984). Prevention of interference by dextran with biuret-type assay of serum proteins. Clin. Chem., 30: 559-561.

Furusawa, E. and Furusawa, S. (1985). Anticancer activity of a natural product, vivo-natural, extracted from *Undaria pinnantifida*. Oncology, 42: 364-369.

Gambrell Jr., R.D., Bagnell, C.A. and Greenblatt, R.B. (1983). Role of estrogens and progesterone in the etiology and prevention of endometrial cancer: Review. Am. J. Obstet. Gynecol., 146(6): 696-707.

Ganju, L., Karan, D., Chanda, S., Srivastava, K.K., Sawhney, R.C. and Selvamurthy, W. (2003). Immunomodulatory effects of agents of plant origin. Biomed. Pharmacother., 57(7): 296-300.

Garg, A., Darokar, M.P., Sundaresan, V., Faridi, U., Luqman, S.R. and Khanuja, S.P.S. (2007). Anticancer activity of some medicinal plants from high altitude evergreen elements of Indian Western Ghats. J. Res. Educ. Indian Med., XIII(3): 1-6.

Gazak, R., Walterova, D. and Kren, V. (2007). Silybin and silymarin: New and emerging applications in medicine. Curr. Med. Chem., 14: 315-338.

George, L. and Eapen, S. (2002). Biotechnological approaches for the development of anticancer drugs from plants. In: *Role of Biotechnology in Medicinal and Aromatic Plants* (*Special Edition on Cancer*), Vol. V, 1st Edn. (Khan, I.A. and Khanum, A., eds.). Ukaaz Publications, Hyderabad, India. pp. 133-162.

Gilman, A.G., Goodman, L.S., Rall, T.W. and Murad, F., eds. (1985). *Goodman and Gilman's The Pharmacological Basis of Therapeutics*, 7th Edn. Macmillan Publishing Co., New York. pp. 1066-1067 and 1115.

Gold, L.S., Stone, T.H. and Ames, B.N. (1997). Prioritization of possible carcinogenic hazards in food. In: *Food Chemical Risk Analysis*. Chapman and Hall, New York. pp. 267-295.

Guo, J.M., Xiao, B.X., Liu, Q., Zhang, S., Liu, D.H. and Gong, Z.H. (2007). Anticancer effect of aloe-emodine on cervical cancer cells involve G_2/M arrest and induction of differentiation. Acta Pharmacol., 28(12): 1991-1995.

Gupta, G.L. and Rana, A.C. (2007). *Withania somnifera* (Ashwagandha): A review. Phcog. Rev., 1(1): 129-136.

Gupta, S.K., Prakash, J. and Srivastava, S. (2002). Validation of traditional claim of Tulsi, *Ocimum sanctum* Linn. as a medicinal plant. Indian J. Exp. Biol., 40: 765-773.

Gupta, V.K. and Sharma, S.K. (2006). Plants as natural antioxidants. Nat. Prod. Rad., 5(4): 326-334.

Gupta, Y.K., Sharma, S.S., Rai, K. and Katiyar, C.K. (2001). Reversal of paclitaxel induced neutropenia by *Withania somnifera* in mice. Indian J. Physiol. Pharmacol., 45: 253-257.

Hallberg, A. and Johansson, O. (2005). FM broadcasting exposure time and malignant melanoma incidence. Electromag. Biol. Med., 24: 1-8.

Haque, R., Bin-Hafeez, B., Ahmad, I., Parvez, S., Pandey, S. and Raisuddin, S. (2001). Protective effects of *Emblica officinalis* Gaertn. in cyclophosphamide treated mice. Hum. Exp. Toxicol., 20(12): 643-650.

Hardell, L., Carlberg, M. and Mild, K.H. (2005). Case-control study on cellular and cordless telephones and the risk for acoustic neuroma or meningioma in patients diagnosed 2000-2003. Neuroepidemiology, 25: 120-128.

Harikrishnan, R., Balasundaram, C. and Heo, M.S. (2010). Herbal supplementation diets on hematology and innate immunity in goldfish against *Aeromonas hydrophila*. Fish Shellfish Immunol., 28(2): 354-361.

Harris, R.B., Griffth, K. and Moon, T.E. (2001). Trends in the incidence of non-melanoma skin cancer is southern eastern Arizona. J. Am. Alad. Dermatol., 45(4): 528-536.

Heber, D. (2004). Vegetables, fruits and phytoestrogens in the prevention of diseases. J. Postgrad. Med., 50: 145-149.

Hecht, S.S., Kenney, P.M., Wang, M., Trushin, N., Agarwal, S., Rao, A.V. and Upadhyaya, P. (1999). Evaluation of butylated hydroxyanisole, myoinositol, curcumin, esculetin, resveratrol and lycopene as inhibitors of benzo[a]pyrene plus 4-(methylnitrosamino)-1-(3-pyridyl)-1-butanoneinduced lung tumorigenesis in A/J mice. Cancer Lett., 137: 123-130.

Helt, A.M. and Galloway, D.A. (2003). Mechanism by which DNA tumour virus oncoproteins target the Rb family of pocket proteins. Carcinogenesis, 24(1): 159-169.

Hertz, R. (1976). The estrogen-cancer hypothesis. Cancer, 38(1): 534-540.

Hirano, T., Oka, K., Mimaki, Y., Kuroda, M. and Sashida, Y. (1996). Potent growth inhibitory activity of a novel ornithogalum cholestane glycoside on human cells: Induction of apoptosis in promyelocytic leukemia HL-60 cells. Cancer Lett., 58: 789-798.

Holinka, C.F., Hata, H., Kuramoto, H. and Gurpide, E. (1986). Effects of steroid hormones and antisteroids on alkaline phosphatase activity in human endometrial cancer cells (ishikawa line). *Cancer Res.*, 46: 2771-2774.

IDMA (2002). *Indian Herbal Pharmacopoeia*. Mumbai, India.

Ichikawa, H., Takada, Y., Shishodia, S., Jayaprakasam, B., Nair, M.G. and Aggarwal, B.B. (2006). Withanolides potentiate apoptosis, inhibit invasion, and abolish osteoclastogenesis through suppression of nuclear factor-kappaB (NF-kappaB) activation and NFkappaB-regulated gene expression. Mol. Cancer Ther., 5: 1434-1445.

Ishimura, K., Matsuda, H., Tatsumi, H., Fujita, H., Terada, N. and Kitamura, Y. (1986). Ultrastructural changes in the ovaries of Sl/Slt mutant mice, showing developmental deficiency of follicles and tubular adenomas. Arch. Histol. Jpn., 49: 379-389.

Jabara, A.G. (1962). Some tissue changes in the dog following stilboestrol administration. Aust. J. Exp. Phil., 40: 293-308.

Jadeja, B.A., Odedra, N.K., Solanki, K.M. and Baraiya, N.M. (2006). Indigenous animal healthcare practices in district Porbandar, Gujrat. Indian J. Trad. Knowledge, 5(2): 253-258.

Jagadeesh, M.C., Sreepriya, M., Bali, G. and Manjulakumari, D. (2009). Biochemical studies on the effect of curcumin and embelin during N-nitrosodiethylamine/ phenobarbital induced-hepatocarcinogenesis in Wistar rats. Afr. J. Biotechnol., 8(18): 4618-4622.

Jagetia, G.C. and Rao, S.K. (2006). Evaluation of the antineoplastic activity of Guduchi (*Tinospora cordifolia*) in Ehrlich ascites carcinoma bearing mice. Biol. Pharm. Bull., 29(3): 460-466.

Jain, J.B., Kumane, S.C. and Bhattacharya, S. (2006). Medicinal flora of Madhya Pradesh and Chhattisgarh: A review. Indian J. Traditional Knowledge, 5: 237-242.

Jain, N.C. (1986). *Schelm Veterinary Haematology*, 4th Edn. Lea and Febiger, Philadelphia. pp. 274-349.

Jarald, E.E. and Jarald, S.E. (2006). *Colour Atlas of Medicinal Plants*. New Delhi, India.

Jayaprakasam, B., Zhang, Y., Seeram, Y.N. and Nair, M. (2003). Growth inhibition of tumor cell lines by withanolides from *Withania somnifera* leaves. Life Sci., 74: 125-132.

Jeena, K., Kuttan, G. and Kuttan, R. (2001). Antitumour activity of *Embilca officinalis*. J. Ethnopharmacol., 71: 65-69.

Jeney, G., Galeotti, M., Volpatt, D., Jeney, Z. and Anderson, D.P. (1997). Prevention of stress in rainbow trout (*Oncorhynchus mykiss*) fed diets containing different doses of glucan. Aquaculture, 154: 1-15.

Jha, N.K. (2007). *Withania somnifera*: Ashwagandha. Phytopharm., 8: 3-35.

Kachmar, J.F. and Moss, D.W. (1976). Enzymes. In: *Fundamental of Clinical Chemistry*. Teitz, N.W., Sounders, Philadelphia. pp. 666-672.

Kader, A.M.M., Hay, A.A., Elwi, A.M. and Eissa, M.H. (1969). Histopathological changes in female albino rats induced by gestogen administration. J. Egypt. Med. Assoc., 52: 359-371.

Kathiresan, K., Boopathy, N.S. and Kavitha, S. (2006). Coastal vegetation: An underexplored source of anticancer drugs. Nat. Prod. Rad., 5(2): 115-119.

Kaur, C. and Kapoor, H.C. (2002). Antioxidants activity and total phenolic content of some Asian vegetables. Int. J. Food Sci. Tech., 37: 153-161.

Kaur, K., Rani, G., Widodo, N., Nagpal, A., Taira, K., Kaul, S.C. and Wadhwa, R. (2004). Evaluation of the antiproliferative and antioxidant activities of leaf extract from *in vivo* and *in vitro* raised Ashwagandha. Food Chem. Toxicol., 42: 2015-2020.

Kaur, S., Michael, H., Arora, S., Harkonen, P.L. and Kumar, S. (2005). The *in vitro* cytotoxic and apoptotic activity of Triphala: An Indian herbal drug. J. Ethnopharmacol., 97(1): 15-20.

Kaushik, P. and Dhiman, A.K. (1999). *Medicinal Plants and Raw Drugs of India*. Bishen Singh Mahendra Pal Singh. Dehra Dun, India.

Kawanishi, M., Sakamoto, M., Ito, A., Kishi, K. and Yagi, T. (2003). Construction of reporter yeasts for mouse aryl hydrocarbon receptor ligand activity. Mutat. Res., 540(1): 99-105.

Khan, K.H. (2009). Roles of *Emblica officinalis* in medicine: A review. Botany Res. Int., 2(4): 218-228.

Khan, M.T., Lampronti, I., Martello, D., Bianchi, N., Jabbar, S., Choudhuri, M.S., Datta, B.K. and Gambari, R. (2002). Identification of pyrogallol as an antiproliferative compound present in extracts from the medicinal plant *Emblica officinalis*: Effects on *in vitro* cell growth of human tumor cell lines. Intl. J. Oncol., 21(1): 187-192.

Kini, D.P., Pandey, S., Shenoy, D.B., Singh, U.V. and Udupa, N. (1997). Antitumour and antifertility activities of plumbagin controlled release formulations. Indian J. Exp. Biol., 35: 374-379.

Koduru, S., Grierson, D.S. and Afolayan, A.J. (2007). Ethnobotanical information of medicinal plants used for treatment of cancer in Eastern Cape Province, South Africa. Curr. Sci., 92: 906-908.

Konczak, I., Okuno, S., Yoshimoto, M. and Yamakawa, O. (2004). Caffeoylquinic acids generated *in vitro* in a high-anthocyanin-accumulating sweet potato cell line. J. Biomed. Biotechnol., 5: 287-292.

Kshirsagar, A., Ingawale, D., Ashok, P. and Vyawahare, N. (2009). Silymarin: A comprehensive review. Phcog. Rev., 3(5): 116-124.

Kumar, S., Suresh, P.K., Vijaybabu, M.R., Arunkumar, A. and Arunakaran, J. (2006a). Anticancer effects of ethanolic neem leaf extract on prostate cancer cell line (PC-3). J. Ethnopharmacol., 105(1-2): 246-250.

Kumar, T., Larokar, Y.K., Iyer, S.K., Kumar, A. and Tripathi, D.K. (2011). Phytochemistry and pharmacological activities of *Silybum marianum*: A Review. Int. J. Pharm. Phytopharmacol. Res., 1(3): 124-133.

Kumar, V., Abbas, A.K. and Fausto, N. (2006b). In: *Pathologic Basis of Disease*, 7th Edn. Saunders, Elsevier India Pvt. Ltd., New Delhi, India. pp. 269-342 and 1059-1117.

Lakra, W.S., Swaminathan, T.R. and Joy, K.P. (2011). Development, characterization, conservation and storage of fish cell lines: A review. Fish Physio. Bioch., 37(1): 1-20.

Lakshmi Prabha, A., Nandagopalan, V. and Kannan, V. (2002). Cancer and plant based anticancer drugs. In: *Role of Biotechnology in Medicinal and Aromatic Plants (Special Edition on Cancer)*, Vol. V, 1st Edn. (Khan, I.A. and Khanum, A., eds.). Ukaaz Publications, Hyderabad, India. pp. 33-67.

Langset, L. (1995). Oxidants, antioxidants and disease prevention. In: *ILSI Europe Concise Monograph Series*. ILSI, Europe/ILSI Press, Brussels.

Lee, J.S., Park, E.H, Choe, J. and Chipman, J.K. (2000). N-methyl-N-nitrosourea (MNU) induces papillary thyroid tumors which lack *ras* gene mutations in the hermaphroditic fish *Rivulus marmoratus*. Teratogen Carcinogen Mutagen, 20: 1-9.

Lee, S.J., Son, K.H., Chang, H.W., Kang, S.S. and Kim, H.P. (1998). Anti-inflammatory activity of *Lonicera japonica*. Phytother. Res., 12: 445-447.

Liehr, J.G. (2001). Genotoxicity of steroidal oestrogens oestrone and oestradial: Possible mechanism of uterine and mammary cancer development. Hum. Repro. Update, 7(3): 273-281.

Lin, C.C. and Huang, P.S.C. (2000). Antioxidant and hepatoprotective effects of *Acanthopanax*. Phytother. Res., 14: 489-494.

Loose, D.S. and Stancel, G.M. (2006). Estrogens and progestins. In: *Goodman and Gilman's The Pharmacological Basis of Therapeutics*, 11th Edn. (Brunton, L.L., ed.). McGraw-Hill Co., New York. pp. 1541-1571.

Madhuri, S. (2008). Studies on oestrogen induced uterine and ovarian carcinogenesis and effect of ProImmu in rat. PhD thesis. RDVV, Jabalpur, MP, India.

Madhuri, S. (2011). Cytogenic effect of *Tinospora cordifolia*, *Withania somnifera* and ProImmu on oestrogen induced uterine and ovarian cytotoxicity in rats. Research project report. CSIR, New Delhi, India.

Madhuri, S. (2012). A perspective on aquaculture: An overview. Life Sci. Bull., 9(1): 111-115.

Madhuri, S. and Pandey, Govind P. (2007). Efficacy of ProImmu on oestrogen induced uterine damage in rat. Int. J. Green Pharm., 2(1): 23-25.

Madhuri, S. and Pandey, Govind (2008). Some dietary agricultural plants with anticancer properties. Pl. Arch., 8(1): 13-16.

Madhuri, S. and Pandey, Govind (2009a). Some anticancer medicinal plants of foreign origin. Curr. Sci., 96(6): 779-783.

Madhuri, S. and Pandey, Govind (2009b). Ethnomedicinal plants for prevention and treatment of tumours. Int. J. Green Pharm., 3(1): 2-5.

Madhuri, S. and Pandey, Govind (2009c). Anticancer activity of *Withania somnifera* Dunal (Ashwagandha). Indian Drugs, 46(8): 603-609.

Madhuri, S. and Pandey, Govind (2010a). Effect of ProImmu, a herbal drug on estrogen caused uterine and ovarian cytotoxicity. Biomed, 5(1): 57-62.

Madhuri, S. and Pandey, Govind (2010b). Experimentally produced ovarian adenocarcinoma by estrogen: A new model. In: *Bioresources for Rural Livelihood: Genetics, Biochemistry and Toxicology*, Vol. I, (Kulkarni, G.K., Pandey, B.N. and Joshi, B.B., eds.). Narendra Publishing House, Delhi, India. pp. 233-237.

Madhuri, S. and Pandey, Govind (2011). Oestrogen induced ovarian hyperplasia leading to cancer in the rat ovary. Int. J. Pharmac. Biol. Arch., 2(2): 639-642.

Madhuri, S., Pandey, Govind, Bhandari, R. and Shrivastav, A.B. (2012a). Fish cancer developed by environmental pollutants. Int. Res. J. Pharm., 3(10): 17-19.

Madhuri, S., Pandey, Govind and Khanna, A. (2007a). Toxicity study of ethinyl oestradiol in female rats. Natl. J. Life Sci., 4(1): 67-70.

Madhuri, S., Pandey, Govind and Khanna, A. (2009). Oestrogen induced uterine damage in rats. Toxicol. Int., 16(1): 5-7.

Madhuri, S., Pandey, Govind, Khanna, A. and Sahni, Y.P. (2007b). Oestrogen induced gynecologic cancer in albino rat. Proceedings: 30-38. National Seminar of 4th MP Science Congress-2007 of the Indian Science Congress Association, Bhopal Chapter held at Govt. Holkar Science College, Indore, MP, India.

Madhuri, S., Pandey, Govind, Khanna, A. and Sahni, Y.P. (2011a). Estrogen induced ovarian damage leading to carcinogenesis in albino rat. Int. Res. J. Pharm., 2(2): 153-156.

Madhuri, S., Pandey, Govind, Khanna, A. and Sahni, Y.P. (2012b). Anticancer effect of a herbal formulation on oestrogen induced ovarian adenocarcinoma in rat. Pl. Arch., 12(1): 353-357.

Madhuri, S., Pandey, Govind, Khanna, A., Shrivastav, A.B. and Quadri, M.A. (2012c). Effect of some herbal drugs on haematological profiles of rats. Int. Res. J. Pharm., 3(12): 158-160.

Madhuri, S., Pandey, Govind, Pandey, S.P. and Jain, V. (2012d). The female hormone oestrogen is not without danger: An overview. Novel Sci. Int. J. Pharmaceu. Sci., 1(6): 386-388.

Madhuri, S., Pandey, Govind and Sahni, Y.P. (2012e). Chemical toxicity leading to cancer in fishes and its treatment by medicinal plants. Pl. Arch., 12(2): 579-584.

Madhuri, S., Pandey, Govind, Sahni, Y.P. and Khanna, A. (2010). Anticancer effect of a herbal drug ProImmu on the experimental uterine cancer in rat. Phytomedica, 11: 51-58.

Madhuri, S., Pandey, Govind, Shrivastav, A.B. and Sahni, Y.P. (2007c). Ovarian cytotoxicity by oestrogen in rat. Toxicol. Int., 14(2): 143-145.

Madhuri, S., Pandey, Govind and Verma, K.S. (2011b). Antioxidant, immuno-modulatory and anticancer activities of *Emblica officinalis*: An overview. Int. Res. J. Pharm., 2(8): 38-42.

Madhuri, S., Shrivastav, A.B. and Pandey, Govind (2012f). Overviews of the zebrafish model and fish neoplasms. The Global J. Pharmaceu. Res., 1(4): 736-743.

Malik, F., Kumar, A., Bhushn, S., Khan, S., Bhatia, A., Suri, K.A., Quzi, G.N. and Singh, J. (2007). Reactive oxygen species generation and mitochondrial dysfunction in the apoptosis cell death of human myeloid leukemia HL-60 cell by a dietary compound withaferin A with concomitant protection by N-acetyl cysteine. Apoptosis, 12: 2115-1233.

Marnewick, J.L., Gelderblom, W.C.A. and Joubert, E. (2000). An investigation on the antimutagenic properties of South African herbal teas. Mutat. Res., 471: 157-166.

Mathew, S. and Kuttan, S. (1999). Immunomodulatory and antitumour activities of *Tinospora cordifolia*. Fitoterapia, 70(1): 35-43.

Mathur, R., Gupta, S.K., Singh, N., Mathur, S., Kochupillai, V. and Velpandian, T. (2006). Evaluation of the effect of *Withania somnifera* root extracts on cell cycle and angiogenesis. J. Ethnopharmacol., 105: 336-341.

Matsui, T., Ebuchi, S., Fukui, K., Matsugano, K., Terahara, N. and Matsumoto, K. (2004). Caffeoylsophorose, a new natural alpha-glucosidase inhibitor from red vinegar by fermented purple-fleshed sweet potato. Biosci. Biotechnol. Biochem., 68: 2239-2246.

Mazumdar, V.K., Gupta, M., Maiti, S. and Mukharjee, D. (1997). Antitumour activity of *Hygrophila spinosa* on Ehrlich ascites aarcinoma and aarcoma-180 induced mice. Indian J. Exp. Biol., 35: 473-477.

Meeran, S.M., Katiyar, S., Elmets, C.A. and Katiyar, S.K. (2006). Silymarin inhibits UV radiation-induced immunosuppression through augmentation of interleukin-12 in mice. Mol. Cancer Ther., 7: 1660-1668.

Meissner, W.A., Sommers, S.C. and Sherman, G. (1957). Endometrial hyperplasia, endometrial carcinoma, and endometriosis produced experimentally by estrogen. Cancer, 10(3): 500-509.

Minervini, F., Giannoccaro, A., Cavallini, A. and Visconti, A. (2005). Investigations on cellular proliferation induced by zearalenone and its derivatives in relation to the estrogenic parameters. Toxicol. Lett., 159: 272-283.

Mishra, L.C., Singh, B.B. and Dagenais, S. (2000). Scientific basis for the therapeutic use of *Withania somnifera* (Ashwagandha): A review. Altern. Med. Rev., 5: 334-346.

Montaser, M., Mahfouz, M.E., El-Shazly, S.A.M., Abdel-Rahman, G.H. and Bakry, S. (2010). Toxicity of heavy metals on fish at Jeddah coast KSA: Metallothionein expression as a biomarker and histopathological study on liver and gills. World J. Fish Marine Sci., 2(3): 174-185.

Moore, M.J. and Stegeman, J.J. (1994). Hepatic neoplasms in winter flounder *Pleuronectes americanus* from Boston Harbor, Massachusetts, USA. Dis. Aquat. Org., 20: 33-48.

Mukherjee, R., Das, P.K. and Ram, G.C. (2005). Immunotherapeutic potential of *Ocimum sanctum* Linn. bovine subclinical mastitis. Rev. Vet. Sci., 79(1): 37-43.

Mukherjee, S., Bidhan, C.K., Ray, S. and Ray, A. (2006). Environmental contaminants in pathogenesis of breast cancer. Indian J. Exp. Biol., 44: 597-617.

Mukhopadhya, A. (2008). Hepatitis C in India. *J. Biosci.,* 33(4): 465-473.

Murdoch, W.J. and Van Kirk, E.A. (2002). Steroid hormonal regulation of proliferative, p53 tumor suppressor, and apoptotic responses of sheep ovarian surface epithelial cells. Mol. Cell Endocrinol., 186: 61-67.

Myers, M.S., Stehr, C.M., Olson, O.P., Johnson, L.L., McCain, B.B., Chan, S.L. and Varanasi, U. (1994). Relationship between toxicopathic hepatic lesions and exposure to chemical contaminants in English sole *Pleuronectes vetulus*, starry flounder *Platichthys stellatus* and white croaker (*Genyonemus lineatus*) from selected marine sites on the Pacific coast, USA. Environ. Health Persp., 102: 200-215.

Nair, S.C., Pannikar, B. and Pannikar, K.R. (1991). Antitumour activity of saffron (*Crocus sativus*). Cancer Lett., 57(2): 109-114.

Nakahata, N., Kutsuwa, M., Kyo, R., Kubo, M., Hayashi, K. and Ohizumi, Y. (1998). Analysis of inhibitory effects of *Scutellaria radix* and baicalein on prostaglandin E_2 production in rat C-6 glioma cells. Am. J. Clin. Med., 26: 311-323.

Nargis, A., Khatun, M. and Talukder, D. (2011). Use of medicinal plants in the remedy of fish diseases. Bangladesh Res. Publ. J., 5(3): 192-195.

Nemmani, K.V.S., Jena, G.B., Dey, C.S., Kaul, C.L. and Ramarao, P. (2002). Cell proliferation and natural killer cell activity by polyherbal formulation, Immu-21 in mice. Indian J. Exp. Biol., 40: 282-287.

Newbold, R.R. and Liehr, J.G. (2000). Induction of uterine adenocarcinoma in CD-1 mice by catechol estrogens. *Cancer Res.,* 60: 235-237.

Norton, W.N. and Gardner, H.S. (2005). Diethylnitrosamine-induced spongiosis hepatis in medaka, *Oryzias latipes*. Microsc. Microanal., 11(Suppl. 2): 1028-1029.

NTP/NIEHS (2001). Advisory panel on federal report on carcinogens makes recommendations to NIEHS/NTP for new listings. National Institute of Environmental Health Sciences press release. www.niehs.nih.gov/oc/news/rocrslt.htm.

Okihiro, M.S. and Hinton, D.E. (1999). Progression of hepatic neoplasia in medaka (*Oryzias latipes*) exposed to diethylnitrosamine. Carcinogenesis, 20: 933-940.

Paiva, S.A. and Russell, R.M. (1999). Beta-carotene and other carotenoids as antioxidants. J. Am. Coll. Nutr., 18: 426-433.

Pandey, Govind P. (1990). Hepatogenic effect of some indigenous drugs on experimental liver damage. PhD thesis. JNKVV, Jabalpur, MP, India.

Pandey, Govind (2007). Role of malnutrition in causing cancer and its safety measures. DNHE project report. Indira Gandhi National Open University, New Delhi, India.

Pandey, Govind (2009). An overview on certain anticancer natural products. J. Pharm. Res., 2(12): 1799-1803.

Pandey, Govind (2010). Malnutrition leading to cancer by some environmental hazards. Int. J. Res. Ayur. Pharm., 1(2): 287-291.

Pandey, Govind (2011a). A review of fish model in experimental pharmacology. Int. Res. J. Pharm., 2(9): 33-36.

Pandey, Govind (2011b). Some important anticancer herbs: A review. Int. Res. J. Pharm., 2(7): 45-52.

Pandey, Govind (2011c). Antioxidant vegetables act against cancer and other diseases. Int. J. Pharmaceu. Stu. Res., 2(1): 32-38.

Pandey, Govind (2011d). Active principles and median lethal dose of *Curcuma longa* Linn. Int. Res. J. Pharm., 2(5): 239-241.

Pandey, Govind (2014). *Remedies of Malnutritional Cancer by Edible Plants*. Daya Publishing House, New Delhi, India.

Pandey, Govind and Madhuri, S. (2006a). Medicinal plants: Better remedy for neoplasm. Indian Drugs, 43(11): 869-874.

Pandey, Govind and Madhuri, S. (2006b). Autochthonous herbal products in the treatment of cancer. Phytomedica, 7: 98-104.

Pandey, Govind and Madhuri, S. (2008a). Median lethal dose, acute and chronic toxicities of ethinyl oestradiol estrogen. Natl. J. Life Sci., 5(2): 291-294.

Pandey, Govind and Madhuri, S. (2008b). Some anticancer agents from plant origin. Pl. Arch., 8(2): 527-532.

Pandey, Govind and Madhuri, S. (2008c). Agriculture plants used against livestock diseases. Int. J. Plant Sci., 3(2): 671-674.

Pandey, Govind and Madhuri, S. (2009a). Some medicinal plants as natural anticancer agents. PHCOG REV., 3(6): 259-263.

Pandey, Govind and Madhuri, S. (2009b). Microbial antibiotics for the treatment of cancers. Drug Invention Today, 1(1): 7-9.

Pandey, Govind and Madhuri, S. (2010a). Global scenario of radiation in cancer risk: A Review. Int. J. Pharmac. Biol. Arch., 1(3): 195-198.

Pandey, Govind and Madhuri, S. (2010b). Cancer pathogenesis caused by xenoestrogens of environment and food contaminants: A review. J. Chem. Pharm. Res., 2(4): 687-695.

Pandey, Govind and Madhuri, S. (2010c). Oncogenic DNA and RNA viruses causing the cancer pathogenesis. Int. J. Pharmaceu. Sci. Rev. Res., 5(1): 120-123.

Pandey, Govind and Madhuri, S. (2010d). Pharmacological activities of *Ocimum sanctum* (Tulsi): A review. Int. J. Pharmaceu. Sci. Rev. Res., 5(1): 61-66.

Pandey, Govind and Madhuri, S. (2010e). Significance of fruits and vegetables in malnutrition cancer. Pl. Arch., 10(2): 517-522.

Pandey, Govind and Madhuri, S. (2011a). Therapeutic approach to cancer by vegetables with antioxidant activity. Int. Res. J. Pharm., 2(1): 10-13.

Pandey, Govind and Madhuri, S. (2011b). Phytoestrogens in cancers and some other disorders. Int. Res. J. Pharm., 2(6): 34-38.

Pandey, Govind, Madhuri, S. and Pandey, S.P. (2009). Ethinyl oestradiol induced liver damage in female albino rats. Indian J. Vet. Pathol., 33(2): 211-212.

Pandey, Govind, Madhuri, S., Pandey, S.P. and Shrivastav, A.B. (2008). Hepatic tissue regeneration by OptiLiv in oestrogen induced hepatotoxicity. Ind. Res. Comm., 2(1): 47-52.

Pandey, Govind, Madhuri, S. and Sahni, Y.P. (2012). Beneficial effects of certain herbal supplements on the health and disease resistance of fish. Novel Sci.: Int. J. Pharmace. Sci., 1(7): 497-500.

Pandey, Govind, Madhuri, S. and Shrivastav, A.B. (2013a). *Recent Research on the Anticancer Herbal Drugs and Cancer Causing Agents.* International E-Publication, ISCA Indore, MP, India.

Pandey, Govind, Madhuri, S. and Shrivastav, A.B. (2013b). Chemical carcinogenesis in fish: A review. Univer. J. Pharm., 2(1): 14-20.

Pandey, Govind, Madhuri, S. and Shrivastav, A.B. (2013c). Prevalence of hepatic and skin cancers in fish by chemical contamination. Int. J. of Pharm. and Research Sci., 2(2): 502-512.

Pandey, Govind, Madhuri, S. and Shrivastav, A.B. (2014). *Fish cancer by environmental pollutants.* Narendra Publishing House, Delhi, India.

Pandey, Govind and Pandey, S.P. (2011). Phytochemical and toxicity study of *Emblica officinalis* (Amla). Int. Res. J. Pharm., 2(3): 270-272.

Pandey, Govind, Pandey, S.P. and Madhuri, S. (2010). Hepatic cell injury by ethinyl oestradiol estrogen. Int. J. Pharmaceu. Stud. Res., 1(1): 49-53.

Pandey, Govind, Pandey, S.P. and Madhuri, S. (2011). Experimental hepatotoxicity produced by ethinyl oestradiol. Toxicol. Int., 18(2): 160-162.

Pandey, Govind and Sahni, Y.P. (2011a). Phytotherapy of malnutritional cancers in animals. Int. J. Pharmaceu. Sci. Rev. Res., 8(1): 14-20.

Pandey, Govind and Sahni, Y.P. (2011b). A review on hepatoprotective activity of silymarin. Int. J. Res. Ayur. Pharm., 2(1): 75-79.

Pandey, Govind, Shrivastav, A.B. and Madhuri, S. (2012). Fishes of Madhya Pradesh with special reference to zebrafish as model organism in biomedical researches. Int. Res. J. Pharm., 3(1): 120-123.

Patel, S., Gheewala, N., Suthar, A. and Shah, A. (2009). *In-vitro* cytotoxicity activity of *Solanum nigrum* extract against *Hela* cell line and *Vero* cell line. Int. J. Pharm. Pharmaceu. Sci., 1(Suppl. 1): 8-10.

Paula, R., Searles, D., Nakanishi, Y., Kim, N.C., Graf, T.N., Oberlies, N.H., Wani, M.C., Wall, M.E., Agarwal, R. and Kroll, D.J. (2005). Milk Thistle and prostate cancer: Differential effects of pure flavonolignans from *Silybum marianum* on antiproliferative end points in human prostate carcinoma cells. *Cancer Res.*, 65: 4448-4457.

Platt, M.E. (2005). Natural Hormone Therapy for Men, Women and Children. www.book1234.net.

Poginsky, B., Westendorf, J., Blomeke, B., Marquardt, H., Hewer, A., Grover, P.L. and Phillips, D.H. (1991). Evaluation of DNA-binding activity of hydroxyanthra-quinones occurring in *Rubia tinctorum* L. Carcinogenesis, 12: 1265-1271.

Polidori, M.C. (2003). Antioxidant micronutrients in the prevention of age-related diseases. J. Postgrad. Med., 49: 229-235.

Poulet, F.M., Wolfe, M.J. and Spitsbergen, J.M. (1994). Naturally occurring orocutaneous papillomas and carcinomas of brown bullheads *Ictalurus nebulosus* in New York State. Vet. Pathol., 31: 8-18.

Prajapati, N.D., Purohit, S.S., Sharma, A.K. and Kumar, T. (2003). *A Hand Book of Medicinal Plants*, 1st Edn. Agrobios (India).

Prakash, J., Gupta, S.K. and Dinda, A.K. (2002). *Withania somnifera* root extract prevents DMBA-induced squamous cell carcinoma of skin in Swiss albino mice. Nutr. Cancer, 42: 91-97.

Premalatha, B. and Rajgopal, G. (2005). Cancer-an ayurvedic perspective. Pharmacol. Res., 51: 19-30.

Premdas, P.D. and Metcalfe, C.D. (1994). Regression, proliferation and development of lip papillomas in wild white suckers, *Catostomus commersoni*, held in the laboratory. Environ. Biol. Fish., 40: 263-269.

Raisuddin, S. and Lee, J.S. (2008). Fish models in impact assessment of carcinogenic potential of environmental chemical pollutants: An appraisal of hermaphroditic mangrove killifish *Kryptolebias marmoratus*. In: *Interdisciplinary Studies on*

Environmental Chemistry- Biological Responses to Chemical Pollutants (Murakami, Y., Nakayama, K., Kitamura, S-I., Iwata, H. and Tanabe, S., eds.). Terrapub. pp. 7-15.

Rajan, B.S. and Chezhiyan, N. (2002). Anticancer drugs from medicinal plants. In: *Role of Biotechnology in Medicinal and Aromatic Plants* (*Special Edition on Cancer*), Vol. V, 1ˢᵗ Edn. (Khan, I.A. and Khanum, A., eds.). Ukaaz Publications, Hyderabad, India. pp. 21-31.

Rajeshkumar, N.V., Pillai, M.R. and Kuttan, R. (2003). Induction of apoptosis in mouse and human carcinoma cell lines by *Emblica officinalis* polyphenols and its effect on chemical carcinogenesis. J. Exp. Clin. Cancer Res., 22(2): 201-212.

Ramasamy, K. and Agarwal, R. (2008). Multitargeted therapy of cancer by silymarin. Cancer Lett., 269(2): 352-362.

Ramnath, V., Kuttan, G. and Kuttan, R. (2002). Antitumour effect of abrin on transplanted tumour in mice. Indian J. Physiol. Pharmacol., 46 (1): 69-77.

Rang, H.P., Dale, M.M., Ritter, J.M. and Moore, P.K. (2003). *Pharmacology*, 5ᵗʰ Edn. Elsevier Science Ltd., New Delhi, India. pp. 693-710.

Rao, K.V.K., Schwartz, S.A., Nair, H.K., Aalinkeel, R., Mahajan, S., Chawda, R. and Nair, M.P.N. (2004). Plant derived product as a source of cellular growth inhibitory phytochemical on PC-3M, DU-145 and LNCaP prostate cancer cell lines. Curr. Sci., 87(11): 1585-1588.

Rao, T.P., Sakaguchi, N., Juneja, L.R., Wada, E. and Yokozawa, T. (2005). Amla (*Emblica officinalis* Gaertn.) extracts reduce oxidative stress in streptozotocin induced diabetic rats. J. Med. Food, 8(3): 362-368.

Ray, G. and Hussan, S.A. (2002). Oxidant, antioxidant and carcinogenesis. Indian J. Exp. Biol., 40: 1213-1232.

Rosangkima, G. and Prasad, S.B. (2004). Antitumour activity of some plants from Meghalaya and Mizoram against murine ascites Dolton's lymphoma. Indian J. Exp. Biol., 42: 981-988.

Rossing, M.A., Cushing-Haugen, K.L., Wicklund, K.G., Doherty J.A. and Weiss, N.S. (2007). Menopausal hormone therapy and risk of epithelial ovarian cancer. *Cancer Epidemiology Biomarkers and Prevention*, 16: 2548-2556.

Sai Ram, M., Neetu, D., Dipti, P., Vandana, M., Ilavazhagan, G., Kumar, D. and Selvamurthy, W. (2003). Cytoprotective activity of Amla (*Emblica officinalis*) against chromium(VI) induced oxidative injury in murine macrophages. Phytother. Res., 17(4): 430-433.

Sancheti, G., Jindal, A., Kumari, R. and Goyal, P.K. (2005). Chemopreventive action of *Emblica officinalis* on skin carcinogenesis in mice. Asian Pac. J. Cancer Prev., 6(2): 197-201.

Sandhu, H.S. (2006). Pesticides as endocrine disruptors. Souvenir and Abstracts: 84-85. International Conference of STOX India held at Gwalior, MP, India.

Satoskar, R.S., Bhandarkar, S.D. and Ainapure, S.S. (2005). Gonadotropins, estrogens and progestins. In: *Pharmacology and Pharmacotherapeutics,* 8th Edn. Popular Prakashan, Mumbai, India. pp. 911-934.

Savlov, E.D., Wittliff, J.L., Hilf, R. and Hall, T.C. (1974). Correlations between certain biochemical properties of breast cancer and response to therapy: A preliminary report. Cancer, 33(2): 303-309.

Schwartz, E., Tornaben, J.A. and Boxill, G.C. (1969). Effects of chronic oral administration of a long acting estrogen quinestrol to dogs. Toxicol. Applied Pharm., 14: 487-494.

Sen, R. (2008). Oral cancer: The global scenario. Proceedings: 1-26 (Part II, Abstracts-Medical Science). 95th Indian Science Congress of the Indian Science Congress Association held at Visakhapatnam, AP, India.

Shar, S.R. and Kew, M.C. (1982). Oral contraceptives and hepatocellular carcinoma. Cancer, 49(1): 407-410.

Sharma, A., Deo, A.D., Riteshkumar, S.T., Chanu, T.I. and Das, A. (2010). Effect of *Withania somnifera* (L. Dunal) root as a feed additive on immunological parameters and disease resistance to *Aeromonas hydrophila* in *Labeo rohita* (Hamilton) fingerlings. Fish Shellfish Immunol., 29(3): 508-512.

Sheng, Y., Pero, R.W., Amiri, A. and Bryngelsson, C. (1998). Induction of apoptosis and inhibition of proliferation in human tumor cells treated with extracts of *Uncaria tomentosa.* Anticancer Res., 18: 3363-3368.

Shi, M., Cai, Q., Yao, L., Mao, Y., Ming, Y. and Ouyang, G. (2006). Antiproliferation and apoptosis induced by curcumin in human ovarian cancer cells. Cell Biol. Int., 30(3): 221-226.

Shrivastava, P.S., Pande, D., Datta, A. and Das, S. (2002). Biotechnological approaches to potential anticancerous herbal drugs of the future. In: *Role of Biotechnology in Medicinal and Aromatic Plants (Special Edition on Cancer),* Vol. V, 1st Edn. (Khan, I.A. and Khanum, A., eds.). Ukaaz Publications, Hyderabad, India. pp. 1-20.

Silva, E.G., Tornos, C., Deavers, M., Kaisman, K., Gray, K. and Gershenson, D. (1998). Induction of epithelial neoplasms in the ovaries of guinea pigs by estrogenic stimulation. *Gynecol. Oncol.,* 71: 240-246.

Singh, M.M., Kambo, V.P., Chowdhury, S.R., Pande, J.K. and Roy, S.K. (1973). Histological and biochemical changes in rat uterus during delayed implantation. Indian J. Exp. Biol., 11: 488-493.

Singh, N., Singh, S.M. and Shrivastava, P. (2005). Effect of *Tinospora cordifolia* on the antitumor activity of tumor-associated macrophages-derived dendritic cells. Immunopharmacol. Immunnotoxicol., 27(1): 1-14.

Singh, R.P., Banerjee, S. and Rao, A.R. (2001). Modulatory influence of *Andrographis paniculata* on mouse hepatic and extrahepatic carcinogen metabolizing enzyme and oxidant status. Phytother Res., 15(5): 382-390.

Sivalokanathan, S., Ilayaraja, M. and Balasubramanium, M.P. (2005). Efficacy of *Terminalia arjuna* (Roxb.) on N-nitrosodiethylamine induced hepatocellular carcinoma in rats. Indian J. Exp. Biol., 43: 264-267.

Somkuwar, A.P. (2003). Studies on anticancer effects of *Ocimum sanctum* and *Withania somnifera* on experimentally induced cancer in mice. PhD thesis. JNKVV, Jabalpur, MP, India.

Srinivasan, S., Ranga, R.S., Burikhanov, R., Han, S.S. and Chendil, D. (2007). Par-4-dependent apoptosis by the dietary compound withaferin A in prostate cancer cells. Cancer Res., 67: 246-253.

Steel, R.G.D. and Torrie, J.H. (1980). Analysis of variance I: The one-way classification/ Multiple comparisons. In: *Principles and procedures of statistics- A Biometrical Approach*, 2nd Edn. McGraw-Hill, Kogakusha Ltd., Tokyo, Japan. pp. 137-194.

Subapriya, R., Bhuvaneswari, V. and Nagini, S. (2005). Ethanolic Neem (*Azadirachta indica*) leaf extract induces apoptosis in the hamster buccal pouch carcinogenesis model by modulation of Bcl-2, Bim, Caspase 8 and Caspase 3. Asian Pac. J. Cancer Prev., 6(4): 515-520.

Sultana, S., Ahmad, S., Khan, N. and Jahangir, T. (2005). Effect of *Emblica officinalis* (Gaertn.) on CCl_4 induced hepatic toxicity and DNA synthesis in Wistar rats. Indian J. Exp. Biol., 43(5): 430-436.

Sumanthran, V.N., Boddul, S., Koppikar, S.J., Dalvi, M., Wele, A., Gaire, V. and Wagh, U.V. (2007). Differential growth inhibitory effects of *W. somnifera* root and *E. officinalis* fruit on CHO cells. Phytother. Res., 21(5): 496-499.

Sur, P., Das, M., Gomes, A., Vedasiromoni, J.R., Sahu, N.P., Banerjee, S., Sharma, R.M. and Ganguly D.K. (2001). *Trigonella foenum-graecum* (Fenugreek) seed extract as an antineoplastic agents. Phytother. Res., 1(3): 257-259.

Takasaki, M., Konoshima, T., Etoh, H., Pal Singh, I., Tokuda, H. and Nishino, H. (2000). Cancer chemopreventive activity of euglobal-G1 from leaves of *Eucalyptus grandis*. Cancer Lett., 155: 61-65.

Tepsuwan, A., Kupradinun, P. and Kusamran, W.R. (2002). Chemopreventive potential of Neem flowers on carcinogen-induced rat mammary and liver carcinogenesis. Asian Pac. J. Cancer Prev., 3(3): 231-238.

Thatte, U.M. and Dahanukar, S.A. (1989). Immunotherapeutic modification of experimental infections by Indian medicinal plants. Phytother. Res., 3: 43-49.

Thompson, L.U., Boucher, B.A., Lui, Z., Cotterchio, M. and Kreiger, N. (2006). Phytoestrogen content of foods consumed in Canada, including isoflavones, lignans and coumestan. Nutr. Cancer, 54(2): 184-201.

Uma Devi, P. (2001). Radioprotective, anticarcinogenic and antioxidant properties of the Indian holy basil, *Ocimum sanctum* (Tulasi). Indian J. Exp. Biol., 39: 185-190.

Uma Devi, P. (1996). *W. somnifera* Dunal (Ashwagandha), potential plant source of a promising drug for cancer chemotherapy and radiosensitization. Indian J. Exp. Biol., 34: 927-932.

Uma Devi, P., Kamath, R.U. and Rao, B.S. (2000). Radiosensitizing effects of a mouse melanoma by withaferin A, *in vivo* studies. Indian J. Exp. Biol., 38: 432-437.

Uozaki, H. and Fukayama, M. (2008). Epstein-Barr virus and gastric carcinoma: Viral carcinogenesis through epigenetic mechanisms. Int. J. Clin. Exp. Pathol., 1: 198-216.

Vecchia, C.L. and Tavani, A. (1998). Fruits, vegetables, and human cancer. Eur. J. Cancer, 7: 3-8.

Vessey, M.P., Lawless, M., McPherson, K. and Yeates, D. (1983). Neoplasia of the cervix uteri and contraception: A possible adverse effect of the pill. Lancet, 2: 930-934.

Vinodhini, R. and Narayanan, M. (2008). Bioaccumulation of heavy metals in organs of fresh water fish *Cyprinus carpio* (Common carp). Int. J. Environ. Sci. Tech., 5(2): 179-182.

Vogel, G. (1977). *New Natural Products and Plant Drugs with Pharmacological, Biological or Therapeutical Activity*. Springer-Verlag, New York. pp. 249-262.

Wang, K., Wu, G. and Dai, S. (2003). Study on the immunological effect of the aqueous extract from Guangxi *Ganoderma leucidum* in the mice. J. Guangxi Med. Univ., 871-874.

Wang, S.Y., Feng, R., Bowman, L., Lu, Y., Ballington, J.R. and Ding, M. (2007). Antioxidant activity of *Vaccinium stamineum*: Exhibition of anticancer capability in Human lung and leukemia cells. Planta medica, 75(5): 451-460.

Wertheimer, N. and Leeper, E. (1979). Electrical wiring configurations and childhood cancer. Am. J. Epidemiol., 109: 273-284.

Wilkinson, J.H., Boutwell, J.H. and Winsten, S. (1969). Evaluation of a new system for the kinetic measurement of serum alkaline phosphatase. Clin. Chem., 15: 487-495.

Xu, M.L., Li, G., Moon, D.C., Lee, C.S., Woo, M.H., Lee, E.S., Jahng, Y., Chang, W.H., Lee, S.H. and Son, J.K. (2006). Cytotoxicity and DNA topoisomerase inhibitory activity of constituents isolated from the fruit of *Evodia officinalis*. Arch. Pharmacol. Res., 29(7): 541-547.

Yin, G., Ardo, L., Jeney, Z., Xu, P. and Jeney, G. (2008). Chinese herbs (*Lonicera japonica* and *Ganoderma lucidum*) enhance non-specific immune response of tilapia, *Oreochromis niloticus*, and protection against *Aeromonas hydrophila*. In: *Diseases in Asian Aquaculture VI, Fish Health Section* (Bondad-Reantaso, M.G., Mohan, C.V., Crumlish, M. and Subasinghe, R.P., eds.). Asian Fisheries Society, Manila, Philippines. pp. 269-282.

Yokozawa, T., Kim, H.Y., Kim, H.J., Okubo, T., Chu, D.C. and Juneja, L.R. (2007). Amla (*Emblica officinalis* Gaertn.) prevents dyslipidaemia and oxidative stress in the ageing process. Br. J. Nutr., 97(6): 1187-1195.

Zhang, Y.J., Nagao, T., Tanaka, T., Yang, C.R., Okabe, H. and Kouno, I. (2004). Antiproliferative activity of the main constituents from *Phyllanthus emblica*. Biol. Pharm. Bull., 27(2): 251-255.

Zhang, Z., Liong, E.C., Lau, T.Y., Leung, K.M., Fung, P.C. and Tipoe, G.L. (2000). Induction of apoptosis by hexamethyl bisacetamide is p53-dependent with telomerase activity but not with terminal differentiation. Int. J. Oncol., 16: 887-892.

Zheng, S., Yang, H., Zhang, S., Wang, X., Yu, L., Lu, J. and Li, J. (1997). Initial study on naturally occurring products from traditional Chinese herbs and vegetables for chemoprevention. J. Cell Biochem. (Suppl.), 27: 106-112.

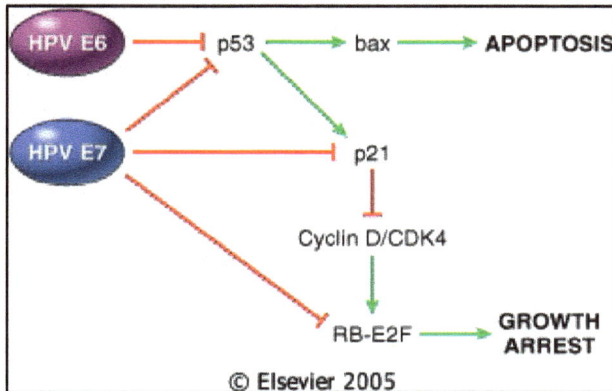

Figure 2: HPV Proteins E6 and E7 Acting on Cell Cycle [Helt and Galloway, 2003; Kumar *et al.*, 2006b]. P. 25

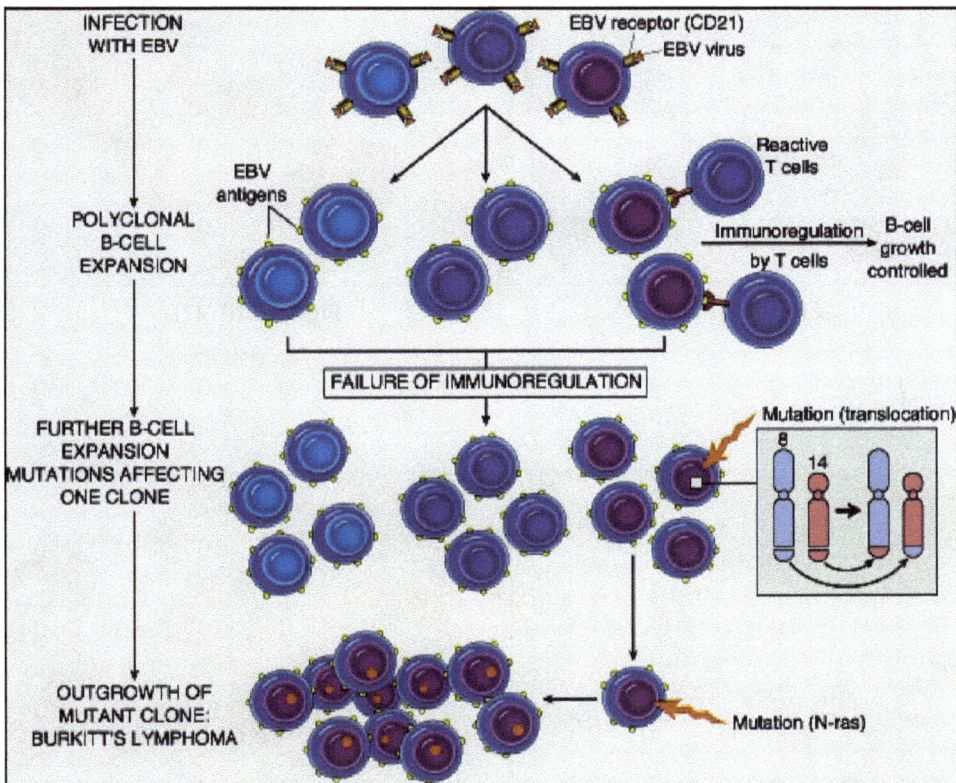

Figure 3: Development of Burkitt Lymphoma by EBV [Dolcetti and Masucci, 2003]. P. 26

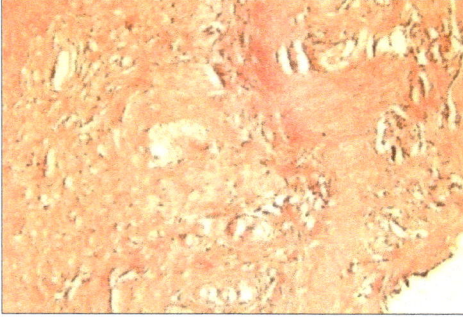

Figure 4 (P. 46)

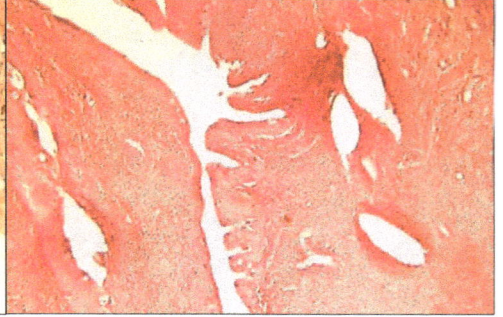

Figure 5 (P. 46)

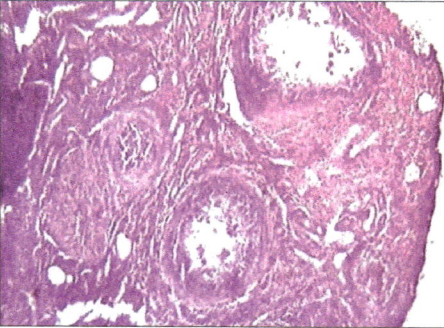

Figure 6 (P. 47)

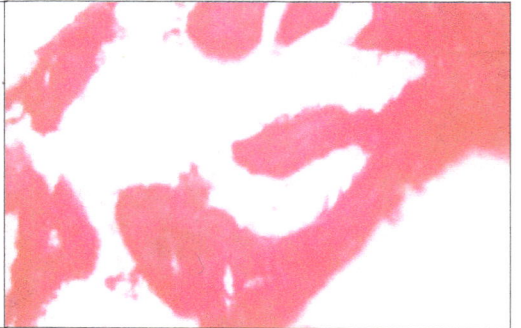

Figure 7 (P. 47)

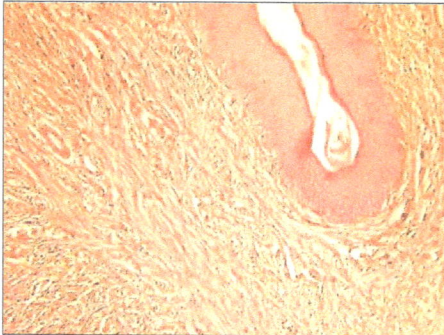

Figure 8 (P. 49)

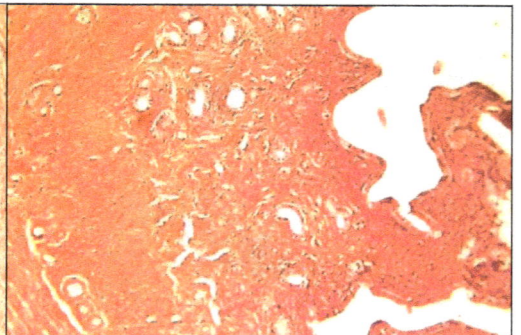

Figure 9 (P. 49)

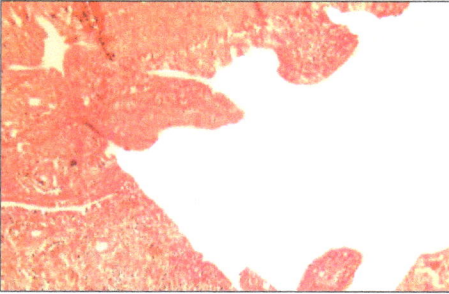
Figure 10 (P. 50)

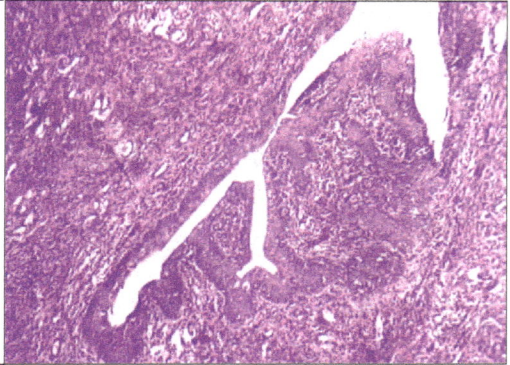
Figure 11 (P. 50)

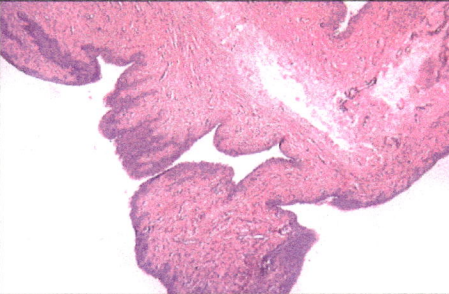
Figure 12 (P. 51)

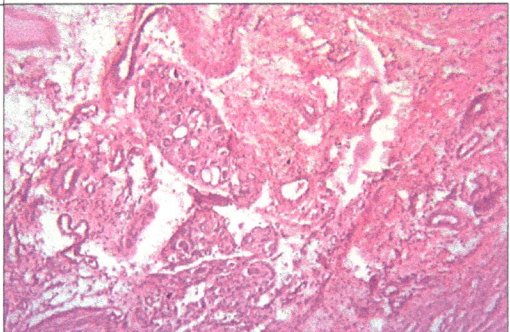
Figure 13 (P. 51)

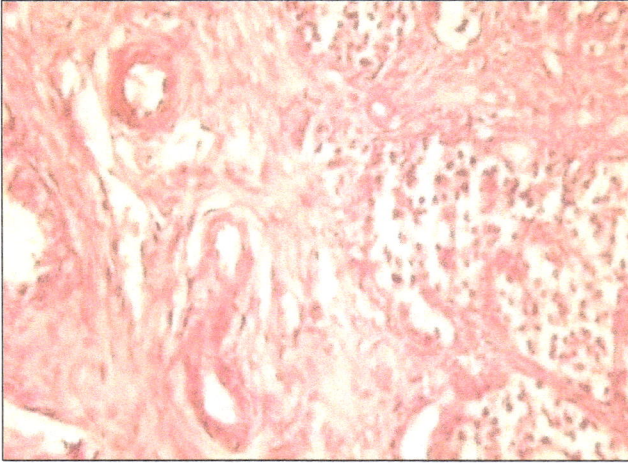

Figure 14 (P. 53)

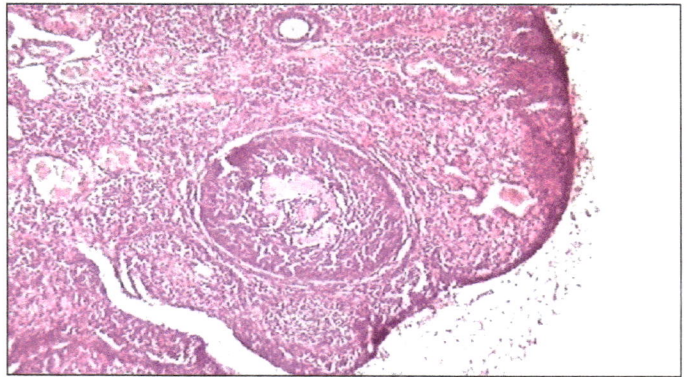

Figure 15 (P. 54)

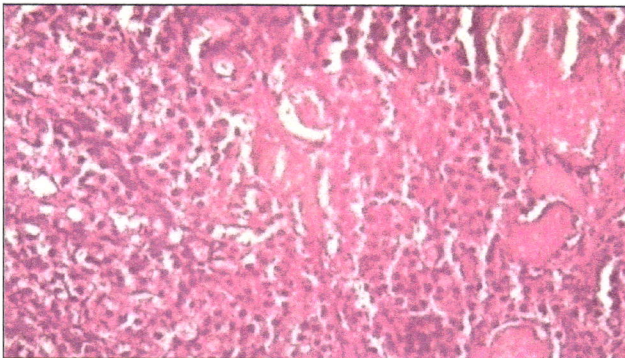

Figure 16 (P. 54)

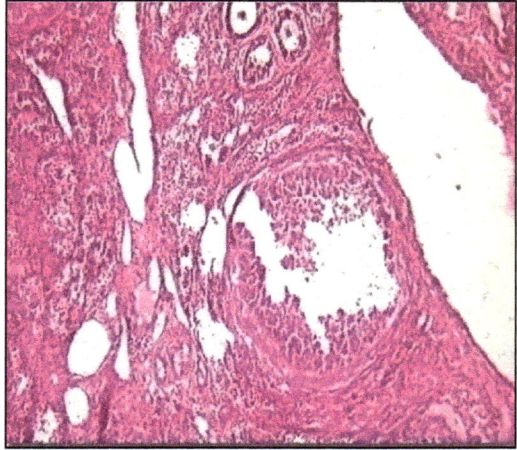
Figure 17 (P. 162)

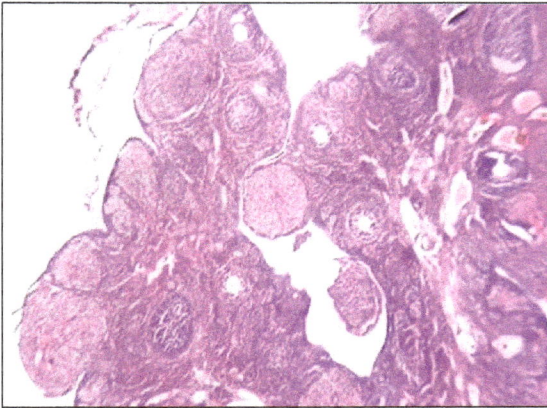
Figure 18 (P. 162)

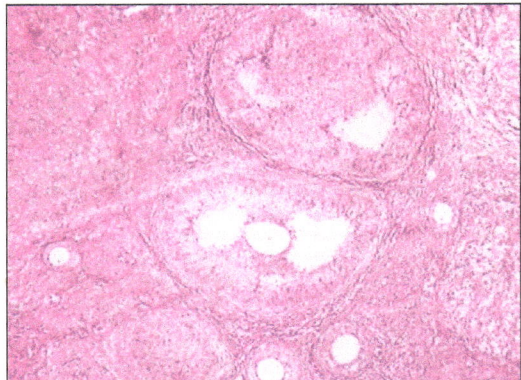
Figure 19 (P. 163)

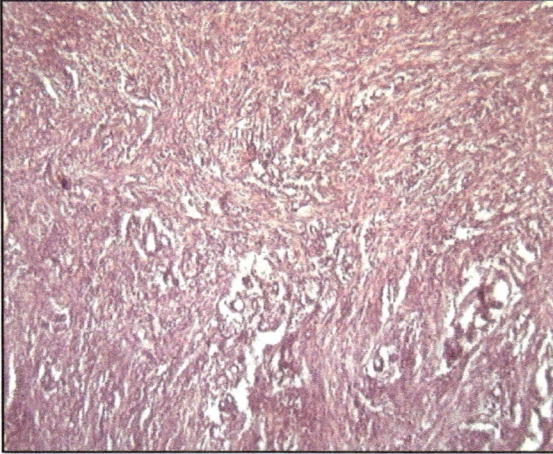

Figure 20 (P. 165)

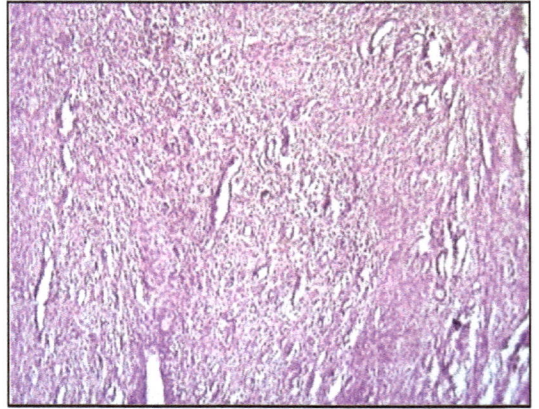

Figure 21 (P. 165)

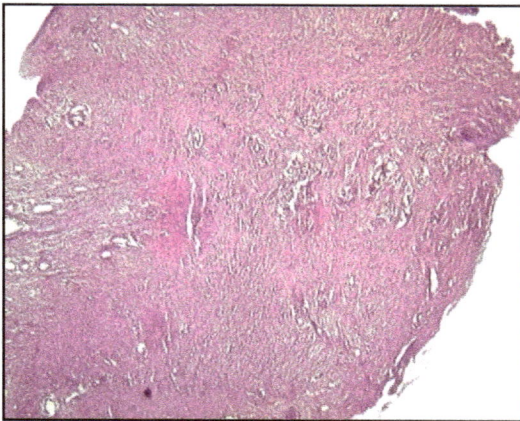

Figure 22 (P. 166)

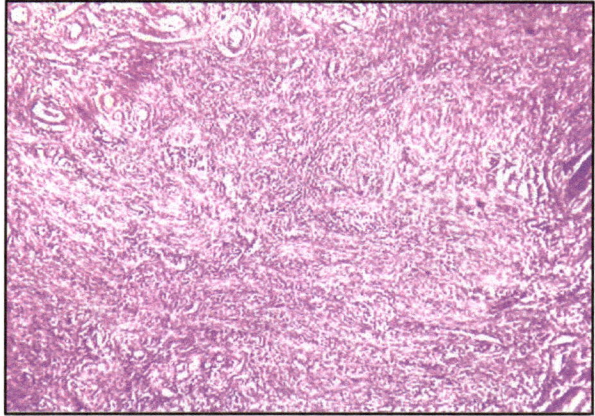

Figure 23 (P. 176)

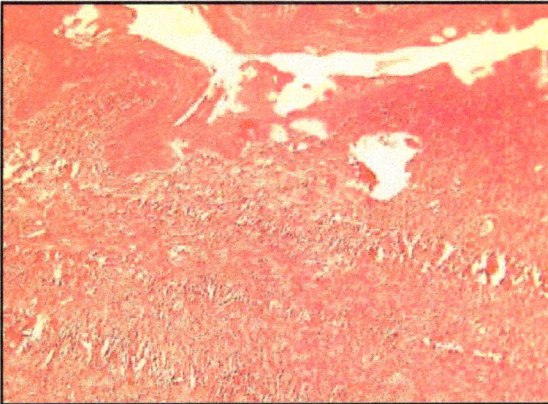

Figure 24 (P. 177)

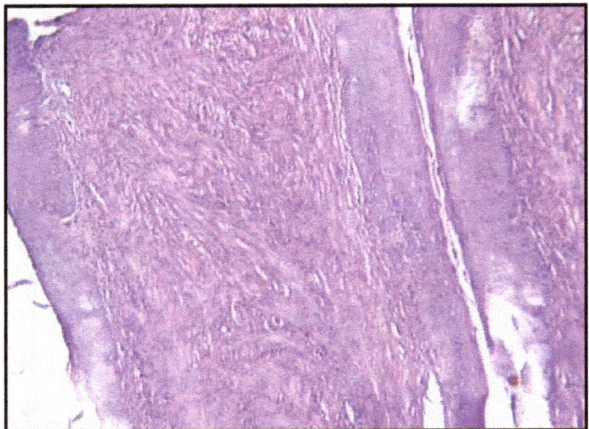

Figure 25 (P. 177)

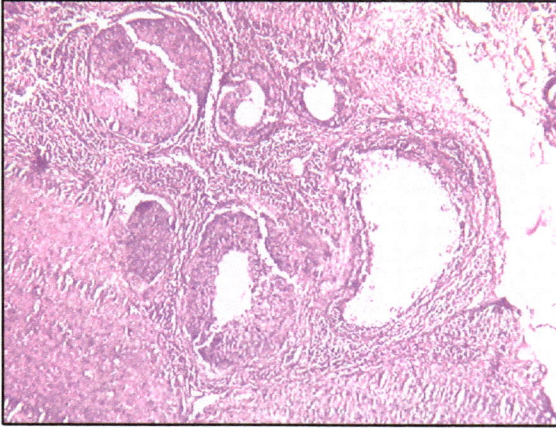

Figure 26 (P. 178)

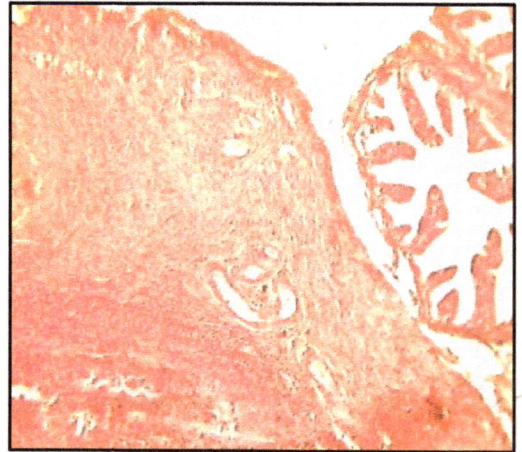

Figure 27 (P. 178)

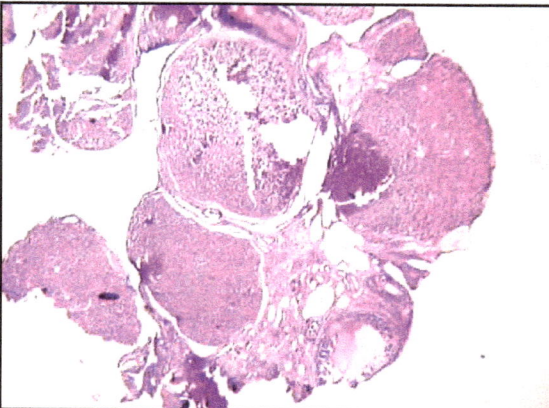

Figure 28 (P. 179)

www.ingramcontent.com/pod-product-compliance
Lightning Source LLC
Chambersburg PA
CBHW050515190326
41458CB00005B/1545